Gregor Mendel's
Experiments on Plant Hybrids

Masterworks of Discovery

Guided Studies of Great Texts in Science

Harvey M. Flaumenhaft, Series Editor

Gregor Mendel's *Experiments on Plant Hybrids*

A Guided Study

Alain F. Corcos
and Floyd V. Monaghan

Original drawings by Maria C. Weber,
Lakewood, Ohio

Rutgers University Press
New Brunswick, New Jersey

"Experiments on Plant Hybrids," by Gregor Mendel. Translated by Eva R. Sherwood for *The Origin of Genetics*, edited by Curt Stern.

Library of Congress Cataloging-in-Publication Data

Mendel, Gregor, 1822–1884.
[Versuche über Pflanzenhybriden. English]
Gregor Mendel's Experiments on plant hybrids : a guided study / Alain F. Corcos and Floyd V. Monaghan : original drawings by Maria C. Weber.
p. cm.—(Masterworks of discovery)
Includes bibliographical references and index.
ISBN 0-8135-1920-9 (cloth)—ISBN 0-8135-1921-7 (pbk.)
1. Hybridization, Vegetable. 2. Plant genetics. I. Corcos, Alain F., 1925– . II. Monaghan, Floyd V., 1916– . III. Title. IV. Series.
QK982.M46 1992
581.1′58—dc20 92-30887
CIP

British Cataloging-in-Publication information available

Manufactured in the United States of America

Contents

Illustrations

Tables

Series Editor's Preface

We often take for granted the terms, the premises, and the methods that prevail in our time and place. We take for granted, as the starting points for our own thinking, the outcomes of a process of thinking by our predecessors.

What happens is something like this: Questions are asked, and answers are given. These answers in turn provoke new questions, with their own answers. The new questions are built from the answers that were given to the old questions, but the old questions are now no longer asked. Foundations get covered over by what is built upon them.

Progress can thus lead to a kind of forgetfulness, making us less thoughtful in some ways than the people whom we go beyond. Hence this series of guidebooks. The purpose of the series is to foster the reading of classic texts in science, including mathematics, so that readers will become more thoughtful by attending to the thinking that is out of sight but still at work in the achievements it has generated.

To be thoughtful human beings—to be thoughtful about what it is that makes us human—we need to read the record of the thinking that has shaped the world around us, and still shapes our minds as well. Scientific thinking is a fundamental part of this record—but a part that is read even less than the rest. It was not always so. Only recently has the prevalent division between the humanities and science come to be taken for granted. At one time, educated people read Euclid and Ptolemy along with Homer and Plato, whereas nowadays readers of Shakespeare and Rousseau rarely read Copernicus and Newton.

Often it is said that this is because books in science, unlike those in the humanities, simply become outdated: in science the past is held to be passé. But if science is essentially progressive, we can understand it only by seeing its progress *as* progress. This means that we ourselves must move through its progressive stages. We must think through the process of thought that has given us what we otherwise would thoughtlessly accept as given. By refusing to be the passive recipients of ready-made presuppositions and approaches, we can avoid becoming their prisoners. Only by actively taking part in discovery—only by engaging in its rediscovery ourselves—can we avoid both blind reaction against and blind submission to the scientific enterprise.

When we combine the scientific quest for the roots of things with the humanistic endeavor to make the dead letter come alive in a thoughtful mind, then the past becomes a living source of wisdom that prepares us for the future—a more solid source of wisdom than vague attempts at being "interdisciplinary," which all too often merely provide an excuse for avoiding the study of scientific thought itself. The love of wisdom in its wholeness requires exploration of the sources of the things we take for granted—and this includes the thinking that has sorted out the various disciplines, making demarcations between fields as well as envisioning what is to be done within them.

Masterworks of Discovery has been developed to help nonspecialists gain access to formative writings in ancient and modern science. The volumes in this series are not books *about* thinkers and their thoughts. They are neither histories nor synopses that can take the place of the original works. The volumes are intended to provide guidance that will help nonspecialists read for themselves the thinkers' own expressions of their thoughts. The volumes are products of a scholarship that is characterized by accessibility rather than originality, so that each guidebook can be read on its own without recourse to surveys of the history of science, or to accounts of the thinkers' lives and times, or to the latest scientific textbooks, or even to other volumes in the Masterworks of Discovery series.

While addressed to an audience that includes scientists as well as scholars in the humanities, the volumes in this series are meant to be readable by any intelligent person who has been exposed to the rudiments of high school science and mathematics. Individual guidebooks present carefully chosen selections of original scientific texts that are fundamental and thought provoking. Parts of books are presented when they will provide the most direct access to the heart of the matters that they treat. What is tacit in the texts is made explicit for readers who would otherwise be bewildered, or sometimes even be unaware that they should be bewildered. The guidebooks provide a generous supply of overviews, outlines, and diagrams. Besides explaining terminology that has fallen out of use or has changed its meaning, they also explain difficulties in the translation of certain terms and sentences. They alert readers to easily overlooked turning points in complicated arguments. They offer suggestions that help to show what is plausible in premises that may seem completely implausible at first glance. Important alternatives that are not considered in the text, but are not explicitly rejected either, are pointed out when this will help the reader think about what the text does explicitly consider. To provoke thought about what are now the accepted teachings in science, the guidebooks bring forward questions about conclusions in the text that otherwise might merely be taken as confirmation of what is now prevailing doctrine.

Readers of these guidebooks will be unlikely to succumb to notions that reduce science to nothing more than an up-to-date body of concepts and facts and that reduce the humanities to frills left over in the world of learning after scientists have done the solid work. By their study of classic texts in science, readers of these guidebooks will be taking part in continuing education at the highest level. The education of a human being requires learning about the process by which the human race obtains its education, and there is no better way to do this than to read the writings of those master students who have been master teachers of the human race. These are the masterworks of discovery.

Discoveries in genetics in recent decades have opened the way to profound investigations into the nature of the human race and to powerful technologies that will radically alter human life. In discussions about what we are and in deliberations about what we are to do, we often hear about genes. We speak of them almost as easily as we speak of eyes or hands. That hasn't always been so. Our easy talk about the gene is the outcome of a process of hard thinking. The origination of that process is commonly attributed to Mendel. But since Mendel himself did not speak of genes at all, we might well ask in what sense Mendel can be called the father of genetics. How was Mendel's work related to the question of inheritance? We cannot answer that question unless we read what Mendel wrote—asking, as we read, just what Mendel's question was and how he answered it.

Mendel's great work was an account of his experiments on plant hybrids. He established empirically that there are constant patterns of disappearance and reappearance of the forms exhibited by the hybrid progeny of peas. Not only did he describe the forms of hybrids and their progeny, he gave a formulation of the patterns in quantitative terms.

Mendel's successors (like his predecessors and his contemporaries) wondered how characteristics are transmitted, with variation and resemblance, from one generation to another. When Mendel's successors considered Mendel's work, they wanted to account for the quantitative patterns that Mendel had shown to be present in hybridization. What sort of means could be responsible for the transmission of characteristics, they asked, given the patterns that Mendel had discovered? Those patterns could result, they answered, only if the material of pollen and egg contains paired particles that are segregated and assorted independently of each other. That question about inheritance was urgent for those who sought a genetic basis for evolution. They drew from Mendel's work a conclusion about the genetic material that Mendel had not drawn. He had not drawn their conclusion because their question was not the one that he had been asking.

If we study Mendel's work under the guidance of Professors Corcos and Monaghan, they will help us to understand Mendel's work in Mendel's own terms, rather than in terms of the answer Mendel's successors gave to their own genetic question. By helping us gain access to the original thinking in a masterwork of discovery, this guidebook to the study of Mendel's *Experiments on Plant Hybrids* can keep us from taking for granted the reworked thought transmitted by Mendel's followers. By helping us to understand Mendel's own thinking, this guidebook can prepare us to understand how Mendel's thought, rediscovered and reworked, entered into the thinking of those who followed him, whom we ourselves now follow.

With minds shaped by the thinking of yesterday and of the day before it, we struggle to answer the questions of today in a world transformed by the minds that did that thinking. We shall proceed more thoughtfully in the days ahead if we have thought through that thinking for ourselves. To help us do so is the purpose of this and all the other guidebooks in the Masterworks of Discovery series.

Harvey Flaumenhaft, Series Editor
St. John's College in Annapolis, June 1992

Preface

At the present time the medical, social, economic, and political implications of genetics are matters of wide public and personal interest and concern. Every so often we hear that genes of economic importance have been transferred from one species to another, that human genes have been introduced into bacteria to permit a hormone, such as insulin, to be produced cheaply and in abundance, that a new and safer vaccine has been developed. These new products result from the use of recombinant DNA techniques whose applications are still in the developmental stage but which promise to revolutionize medical treatment, raise food production to new levels, and give new approaches to forensic evidence in court. DNA technology, like any other technology, raises many questions about its appropriate use and safety. Because the science of genetics is now revolutionizing our life in many ways, it seems appropriate to explore its roots.

The science of genetics dates from 1900 when three botanists rediscovered a paper on plant hybrids written in 1866 by Gregor Mendel. Mendel's paper attracted little attention at the time he wrote it. After its rediscovery in 1900, however, it was seen as a major turning point for the science of biology. It came to be regarded as the founding document of the science of genetics because it was interpreted—incorrectly as we shall see—as providing the first particulate theory of heredity based on real experimental evidence and as containing the first two fundamental laws of heredity: the law of segregation of genes and the law of independent assortment of genes.

After 1900 the science of genetics developed very rapidly, and new discoveries pushed the original paper into the background. As a result, the paper has attracted less attention than it merits. Most geneticists have little time or inclination to delve into the history of their discipline because they are concerned with the newest research. In fact, very few people have read Mendel's paper and, among those who have, very few have understood it.

Recently, however, there has been a resurgence of interest in Mendel among historians of science, in particular those with ties to a special research center in Czechoslovakia. This center, called the Mendelianum, is a part of the Moravian Museum and is located in the former Augustinian monastery in Brno where Mendel lived and worked. In the last twenty-five years, as a result of careful historical research and critical examination of Mendel's paper, it has become apparent that the life and work of Mendel as they are described in textbooks of biology and general biographies of Mendel are often more mythical than factual.[1]

A close study of his paper reveals that the laws of heredity, which are supposed to be there, are not present. Instead, one finds a series of laws relating to the formation of hybrids, which are entirely different from the traditional "Mendelian" laws of heredity. And instead of a theory of heredity, one finds a theory of formation of hybrids. In fact, the word *heredity* occurs only once in the entire text and then in relation to something that does not happen.[2] In addition, there is no mention of anything that corresponds to the concept of a gene, even in an embryonic form.

Why Mendel's paper was seen as relating to heredity when Mendel himself saw it as relating to hybrids is an interesting question. The most likely answer is the following: In the formation of hybrids, characters must be passed from the parent plants to their offspring. Since this is basic to inheritance, Mendel's data could be interpreted as being about heredity. But, we want to repeat, the data were not interpreted in that way by him.

If Mendel is not to be considered the founder of genetics, you may well ask why his work should be included in this series on great texts in science. The answer is very simple. Mendel was a great scientist, a pioneer in integrating ideas across three scientific disciplines—botany, physics, and mathematics. Through this integration he created a mathematically precise quantitative theory of the formation of hybrids and the development of their offspring through several generations. By means of this theory he was able to explain some behaviors of hybrids known for a long time but not then understood. Furthermore, Mendel was a brilliant experimenter. His experiments with hybrids were so well conceived and so well carried out that they remained unchallenged even when they became the basis of the new science of heredity.

The quality of his experimental work is perhaps best shown by the fact that his experiments, and not those of his so-called rediscoverers, are the ones quoted in textbooks of biology and genetics. The magnitude of his creative power is clearly shown by the fact that, starting with a single clue, he carried through the process of developing a large series of experiments, did the experiments over an eight-year period, discovered fundamental laws, and created a theory that provided a satisfactory explanation of his results. It is very rare for one individual to carry out the whole process of science so completely and so elegantly.

We believe that it is now time for Mendel to be better known and understood. Therefore, we have written an interpretive text based upon his original paper. For this purpose, the text has been divided into sections. Each section begins with a brief introduction, followed by Mendel's text, and then the relevant interpretive comments, which are keyed to the numbered lines of each section in the text. Notes at the ends of chapters refer to matters not covered in the introduction or the interpretive text; footnotes are Mendel's.

The book also contains a short biography of Gregor Mendel. In it we have traced, so far as they can be discovered, the influences that shaped his development from childhood through his school years to his mature years as a university student, teacher, research scientist, and, finally, abbot of the Augustinian monastery of St. Thomas. This monastic community played a very important role in the educational, social, and cultural life of Moravia, then a part of the Hapsburg Empire. Many of the members of this community were university graduates. Among them were scientists, philosophers, mathematicians, and musicians. Nearly all of them were teachers, as was Mendel. Just as Mendel's peas flowered and set seeds, leading him to develop his theory of hybrids, his many talents flowered in this stimulating atmosphere. In this biography we have painted a very different picture of Mendel than the familiar one of a cloistered monk puttering around in his pea garden. Here you will find him very involved in the religious life of the monastery, in scientific and agricultural societies, and in business, finance, and community affairs.

We have also included notes on plant breeding and reproduction in flowering plants, with special reference to edible peas. Finally, we have included a glossary and a bibliography.

This interpretation of Mendel's text could not have been written by either of the authors alone, but is the result of their extensive collaboration during the last ten years. It is the evolutionary product of long, ardent discussions in which ideas were born, developed, and matured to the point where neither author can remember which one of them thought of the idea first.

We hope that with such an interpretive text the reader will be able to follow the way Mendel decided on his problem, how he delimited it, why he chose his experimental material, how he obtained his results, and how he analyzed and interpreted them. In short, we hope that the text will offer the reader insight into the way a great scientist practiced his or her science.

The authors express their thanks to the following persons: Gabriel Gogola, Berkley, Michigan, and Paul Forsthoefel, S.J., Professor Emeritus of Biology at the University of Detroit, for their translation of Mendel's two sermons.

To Dr. Petr Fischer, Professor Emeritus of Humanities and Religious Studies at Michigan State University, for his interpretation of Mendel's sermons and of a Czech statement prepared by the late Dr. Alexander Heidler for the commemoration of the publication of Mendel's research.

To their daughters, Christine Corcos and Sheila and Susan Monaghan, and their friends, Steve Hack, Wayne and Nancy Claflin, Carl Thompson, Edward Graham, for their patience in reading the first draft of the book; to Henry Silverman and Lois Zimring for their criticisms of the second draft of the book. Many of their comments and criticisms have been incorporated in the final draft of the book.

To Drs. Vitezlav Orel and Anna Matalova of the Mendelianum Museum at Brno, Czechoslovakia, for welcoming so warmly one of the authors (A.C.) in September 1991, for sharing wih him their ideas and helping him take many photographs.

The writing of this book has been generously supported by a grant from the National Endowment for the Humanities.

Notes

1. We are using the term *mythical* in the sense that the historian of science K. Merton was using it, "to denote a set of ill-founded beliefs held uncritically

by an interested group" K. Merton, "Behavior Patterns of Scientists," *American Scientist* 57 (1969): 1–23.

2. In the section where Mendel discussed the first generation of offspring from his monohybrids he said: "In individual seeds of some plants the green coloration of the albumen is less developed, and can be easily overlooked at first. The cause of partial disappearance of the green coloration has no connection with the hybrid character of the plants, since it occurs also in the parental plant; furthermore this peculiarity is restricted to the individual and not *inherited* [italics added] by the offspring."

The Background

Portrait of Gregor Mendel, ca. 1868 (photo by Mrs. A. F. Corcos)

Life of Mendel

Gregor Mendel, whom the world would one day remember as the "father of the science of genetics," was born on July 20, 1822, and christened Johann two days later. Gregor was his religious name, chosen when he took his vows at the age of twenty-one and became an Augustinian monk at the monastery of St. Thomas in Brünn. Johann was the only son of Anton and Rosina (Schwirtlich) Mendel, who were married October 6, 1818. He had an older sister, Veronica, and a younger one, Theresia.

For more than 130 years the Mendel family had lived in the village of Heinzendorf in Lower Silesia. The village is now called Hyncice and is part of the town of Odry in Czechoslovakia. Johann's direct ancestors had lived on the same farm where he was born. Both Lower Silesia and Moravia were then part of the Hapsburg Empire, a collection of very diverse territories extending from the spurs of the Alps in the north to the shores of the Adriatic in the south and from the Bavarian frontier in the west to the Hungarian plains in the east. Those lands were not bound together either by geography or by a common language, but had been joined by Hapsburg military conquests and political marriages. Although the Hapsburgs were themselves German in origin and the official language of the empire was German, their empire comprised a great variety of peoples, whose diversity of languages and cultures frequently caused strife not only among the districts themselves but also with the central government. Calls for independence were often heard and uprisings were not uncommon.

Since the inhabitants of the district that included Heinzendorf were mainly Germans, German was the common language of the people in the village and thus the language Johann spoke as he was growing up. As we shall see, this had important consequences for him, beginning with his elementary school education and ending with his election as abbot of the monastery of St. Thomas at Brünn. In the elementary schools of each district in the Hapsburg Empire, classes were conducted in the native language of that district, but all education beyond that level was carried on in German. German was not only the language in which the business of government was carried on, but was also the language of commerce, of science, and of literature. For someone growing up in Moravia, where the local language was Czech, it was thus necessary to learn German in order to go to a higher school or a university. For this reason, the simple fact that Mendel's native language was German gave him many advantages. However, when he became a priest and lived in Brünn, which was and still is in Moravia, he had to learn Czech in order to talk to many of the people in his parish and to deliver sermons.

Johann was also fortunate in having the father and mother he did. His father, Anton, was born on April 10, 1789, the second son of Valentin Mendel. We do not know very much about Anton, but we do know that he served for eight years in the Austrian army during the latter part of the Napoleonic Wars, which effectively ended with Napoleon's defeat at Waterloo in 1815. Anton was then discharged and returned to Heinzendorf. Simple arithmetic tells us that at that time he would have been about twenty-six years of age. We do know that when he returned home he came with new ideas.

One evidence of these new ideas was his desire for a house different from and better than the one that had satisfied his ancestors for several generations. As soon as he had taken over the management of the farm, he tore down the old wooden house and replaced it with a solid new one, unlike all others in the village. It was built of stone and had a tile roof, suggesting that he may have served at least part of his eight years in the army in northern Italy, where this type of house was common. A large part of the Austrian army was stationed in northern Italy around 1809 because disturbances occurred frequently in that part of the Hapsburg Empire. If he did serve in Italy, Anton Mendel would have been exposed to a very different culture and a very different way of living.

Another of Anton's new ideas was to plant a large orchard of fruit trees on his farm. He was encouraged in this by the village priest, Father Schreiber, who advised him to improve the orchard by replacing old varieties of fruits with better ones. One way of doing this is by **grafting**. In this procedure one inserts a shoot (called a **scion**) of the

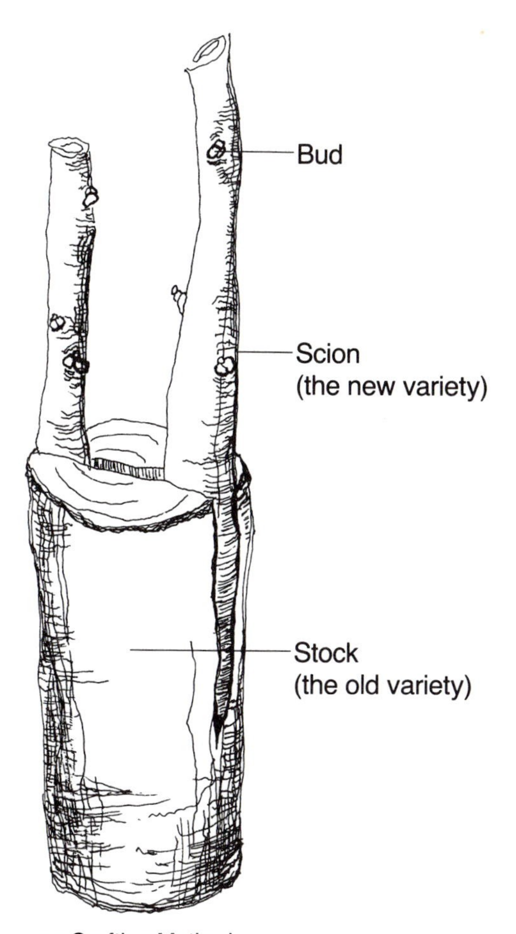

Fig. 1.1. A Common Grafting Method

desired variety into a cut in the trunk, or **stock**, of the tree of the discarded variety (figure 1.1). If there is close contact between the shoot and the trunk, the shoot will bond to the trunk and grow—we say then that the graft has taken. The result is a tree that bears fruits of the desired variety, although the roots of the tree are those of the old variety. Many of the scions that Father Schreiber brought to Anton came from trees on the estate of Countess Maria Truchsess-Ziel, who was the feudal proprietor of the land on which Mendel's father and mother had their farm. Many of these trees were exceptionally fine varieties that the countess had introduced from foreign lands. She gladly gave these materials to Father Schreiber, knowing that they would be well used in improving orchards throughout the district. Her gifts to Father Schreiber were only one of her ways of encouraging the peasants to improve their orchards and gardens. Being wise in the ways of humans, she warned the peasants not to steal grafting material from her gardens and orchards on pain of punishment. She knew this would offer them an irresistible challenge to do exactly what they were forbidden to do. And that is exactly what they did, with exactly the results she desired—better gardens and better orchards throughout the district.

The boy Johann was happy helping his father making grafts to improve the family orchard. Grafting, such as they were doing, is an exacting art and must be done with skill and meticulous attention to detail if the grafts are to take. Johann could see the beautiful results he and his father obtained when the process was done properly and the ugly dead stubs that resulted when it was not. This must have made a large and long-lasting impression on Johann, for in later years, when he was doing his own research on peas and other plants, he was meticulous both in his procedures and in his record keeping. Had he not been meticulous, he never could have obtained the results he did. In addition to contributing to his son's interest in experimenting with making changes in growing things, Anton also set him a good example by working hard and being careful with his income. These habits were of vital importance to Johann, not only in his struggles to get an education, but also later in his management of the monastery at Brünn when he became its abbot.

Johann Mendel was also fortunate in having Rosine as his mother. She was the daughter of Martin Schwirtlich, a professional gardener, one of a family who had been gardeners on the large estates of the nobility for several generations. However, not all the men of the Schwirtlich family were gardeners. One of Rosine's uncles managed to educate himself to a level such that, when discharged from the army in the late 1770s, he was able to establish and run an elementary school for the village children of Heinzendorf. He maintained the school from 1780 to 1788. It was easy for the villagers to see that having a school in their own village was much better for their children than having them go through all kinds of weather to a neighboring village to learn to read and write.

After Anton Schwirtlich's death the school was discontinued for lack of a teacher. However, the benefits of having a local school had impressed the villagers enough for them to apply to the government to have a public grammar school established in the village. The school was finally established in 1796 with Thomas Makitta as its first teacher. It was in this school that Johann got his first formal education. To use an old expression, he took to education like a duck to water, and Makitta was soon aware that Johann was an exceptional student.

We have noted thus far how events beyond Johann's control gave him special advantages. The same was true in Mendel's early schooling. Because this school was in a district where the Countess Truchsess-Ziel had a great deal of influence, the curriculum included not only the basic subjects of reading, writing, and arithmetic, but also, at her insistence, such natural history and natural science as were appropriate for elementary school children. And in a garden near the school the children were taught the basic skills of growing various fruits and keeping bees. Makitta was supported and helped in these matters by Father Schreiber. It was soon evident to both of them that in those subjects Johann was also an exceptional student. This led these two good men to talk to Rosine and Anton Mendel about finding some way for Johann to continue his studies when he had finished with grammar school.

Young Scholar

The only way Johann's parents could continue his education was to send him to a neighboring town where an upper elementary school was available. To do so they had to be able to pay his board and room away from home, the fees for his classroom instruction, and the cost of traveling to and from school. Such costs imposed real hardship on the Mendels. Anton was still required to spend three days out of each week working on the estate of Countess Truchsess-Ziel. This required labor was a last remnant of the feudal labor dues a peasant or a serf was required to pay to the landlord. These labor requirements took up so much of Anton's time that he had only a modest income from his farm and so was unable to contribute much to Johann's expenses.

The decision to send Johann away to school was not an easy one for Anton to make. Since Johann was his only son, Anton had expected that in due time Johann would take over the family farm as he, Anton, had and as his father had before him. However, if Johann did not take over the farm, it would have to be sold. This would mean giving up a landholding that had been in the family for more than a century. All of Anton's plans and hopes seemed to be coming to a dead end. As we shall see later, they very nearly did, but for another reason entirely.

The decision to leave the farm and his parents in order to continue his studies was also a difficult one for Johann. He loved his father and mother dearly, as we know from his letters, especially those to his mother. Although he loved the farm and farming, he could see that his father was leading a hard, grinding, narrow kind of life, which had aged him prematurely. The contrast between this life and the life led by educated people such as Thomas Makitta and Father Schreiber was clear enough to help Johann decide in favor of going on to a higher school.

Probably, the person whose feelings did the most to resolve these conflicting demands and desires was Johann's mother, Rosine. Father Schreiber and Thomas Makitta had almost certainly urged upon her the importance of Johann's continued education. Undoubtedly, these two men felt that an intelligence such as Johann's ought not to be lost, as it would be if he did not continue his education but became a peasant farmer like his father. Certainly his mother wanted a better life for her son than that which his father, Anton, had had, and she understood that further education would give Johann the best chance to have that better life. The best way these three friends could devise for Johann to continue his studies and to test their estimate of his ability was to send him to the school at the nearby town of Lipnik and enroll him at the upper-level grammnar school run by the Piarist Fathers. The Fathers offered their teaching services as part of their religious duties and did not charge any fees. This would make it easier for the Mendels to meet the cost of sending Johann there. In addition, there were already two other boys from Heinzendorf at the school, so he would not be entirely alone among total strangers this first time that he was away from his home and family. He was then eleven years old.

Thus, with mingled hopes, anticipations, heartaches, and longings, Johann went off to Lipnik for a year to find out how he would fare. No one need have had any doubts, for he found himself in his natural element. His achievements must have gladdened his mother and given even Anton a feeling, perhaps reluctant, of pride. Johann's "report card" lists him as a student in the third class and shows him as very diligent and as standing at the very top of his class, a pattern he was to repeat later at other schools. Clearly, he had taken his father's example of working hard and paying attention to details to heart. Both Makitta and Schreiber must also have been pleased that their judgment had been confirmed. We don't know how Johann felt, but the chances are that he enjoyed his studies and was quietly happy that he had been able to justify the faith of his supporters at home and give them the feeling that their hard-earned money had not been wasted. His achievements clearly pointed beyond Lipnik to further education. The school at Lipnik was a technical school with a curriculum designed to train boys for "the arts, the sciences, and commerce" (Iltis [1932] 1966, 33) not to prepare them for higher education or the priesthood. If Johann was to go on to a university or to become a priest, he needed a different kind of preparation than he could obtain at the school in Lipnik. To get this he needed to go to a *gymnasium.*[1]

Johann was to mount that next step up the ladder at the gymnasium at Troppau, now called Opava, a town still farther from home than Lipnik. He went there the next year, 1834, when he was twelve. The school, a classical high school, brought Johann into contact, for the first time, with boys whose backgrounds were very different from his own. He was the son of a peasant family, while his new schoolmates were mainly from upper middle-class families. Johann's cultural deficiencies in background and in social graces and his poor clothing marked him among his fellow students as just what they might expect a peasant's son to be. Probably, in the normal manner of boys of that age and social position, they sneered at Johann and taunted him for his differences. He was clearly a disadvantaged minority facing stinging challenges to his self-esteem and sense of self-worth. The one place in which he could demonstrate his superiority was in the classrooms of the gymnasium. There the peasant's son could give back as good as he got. He could and did surpass most of his fellow students in every class in which he was enrolled. This was quite remarkable because, in order to stay in school, he not only had to keep up with his classroom studies, but also had to find work outside of school to help pay his fees and buy food. Although his parents were better off than most of the other villagers in Heinzendorf, they still could not afford to cover all of Johann's expenses. But whenever someone was going from Heinzendorf to Troppau they sent him whatever food they could spare. Johann could get along fairly well most of the time, though occasionally, on the days when he had little to eat, he felt the bite of hunger pains. Nonetheless, being Anton's son, he accepted the hardships and deprivations of his life and went forward with his schoolwork.

His interest in the natural sciences, which had started in his childhood experiences with his father and also with Father Schreiber, was reinforced by his contacts with the teachers at the gymnasium. Many of them were actively involved in building up the natural history collections at the local museum. We also know that some of them were interested in meteorology and were making quite good observations of the weather elements. Later in his life, Johann also devoted a great deal of time to making and reporting this type of observation. In fact, he became a well known and respected professional meteorologist.

The curriculum of the gymnasium required six years to complete. The first four years were devoted to what were called grammar classes, and the last two were devoted to the upper or *humaniora* classes (the humanities). Johann was in the lower classes from 1834 to 1838, that is, from the time he was twelve until he was sixteen. In his autobiography he wrote that in 1838 his parents suffered many reverses. Some of these may have been crop failures, sick farm animals, or illnesses among the members of his family. Johann was not specific about what they may have been. We do know about one major disaster during the winter of 1838. While doing his required labor for the countess, Johann's father was struck by a falling tree and his chest was partly crushed. Although he was not killed, the injury was so severe that from that day forward he was unable to do any hard physical labor. This effectively put an end to the meager support that he and his wife had been able to send Johann. Thus, when he was sixteen, Johann was on his own, having, as he said in his autobiography, "to provide for himself entirely" (Iltis 1954, 231)

Johann's autobiography does not tell us anything of how he felt about his father's injury, but we know from his later experiences as a young priest that he was extremely sensitive to the suffering of those who were ill or injured and in pain. Knowing this, we can be sure that his father's injury must have affected him deeply, especially since he could not do anything to relieve his father's pain. Imagine, if you will, the thoughts and feelings going through Johann's mind at this point. What should he do, leave the gymnasium, go home, and take over the farm, or stay at the gymnasium and try his best to finish the rest of that year and the remaining two? Going back to the farm meant abandoning his studies. Staying at the gymnasium meant abandoning his father and mother in a time of great need, probably forcing them to sell the farm, giving up the house and everything that it represented, all the family memories for over a century. It meant avoiding spending the rest of his life as a peasant farmer, whose hard lot he knew intimately, and it also meant continuing with the studies he loved so well and which offered the only means he could see for gaining a better life. We do not know whether Johann wrote to his mother about his difficult choice or whether he went home and talked with her. In either case, it seems probable that she would have urged him to continue his studies, not to throw away all that he and they had struggled so hard to gain. Whether he consulted her or not, that was the choice Johann made. He decided to continue with his studies.

Because he needed, as he said, to take care of himself and support himself, he took courses designed to train him to become a private tutor. At the end of these courses, which he apparently took while keeping up with his regular gymnasium classes, he wrote that "following his examination, he was highly recommended in the qualification report," so was able to earn "a scanty livelihood" through private tutoring (Iltis 1954, 231). At the end of his year's work, in the spring of 1839, he most likely went home to Heinzendorf to work on the farm, probably to help his brother-in-law, Alois Sturm, the husband of his older sister, Veronica.

In the fall of 1839 he returned to Troppau, began the studies of his first upper-level or humanities year, and again took up his tutoring. We must suppose that he also worked at such other odd jobs as he could find to help provide food, shelter, clothing, and his fees at the gymnasium. It is probable, given his summer of working on the farm, that he was already more than a little tired when he returned to school. However, he kept up his demanding routine until late May or early June. Then he became so seriously ill that he had to give up his studies temporarily and go home. We can imagine that this required his mother to divide her energies between Anton and Johann, doing her best to restore each of them to some degree of health.

Her success with Anton was, as one would expect, limited, but her results with Johann were better. In September he returned to Troppau and, on the basis of his results up to the point where he had left, was admitted to the second humanities year. He finished his studies at Troppau on August 7, 1840, graduating once again with very high honors—having earned first place in all his examinations. By this time his fellow students must have been acutely aware that although he was a peasant's son he was no ordinary clodhopper.

To complete his preparation for higher education, Johann still needed two years of training in philosophy. To obtain this training he had to leave Silesia and go to Olmütz (now Olomouc) in Moravia. There he applied for admission to the Olmütz Philosophical Institute which was associated with the University of Olmütz, and was accepted. Training in pure philosophy was only a part of the philosophy curriculum, which also included required courses in theology, Latin literature, and the fundamentals of mathematics and physics. Johann became especially interested in these last two subjects.

However, there was also another force at work here shaping Johann's future. The teachers at the institute, as at the gymnasium at Troppau, were mainly members of the Augustinian order. As we shall soon see, his contacts with the Augustinians, and with one of them in particular, Dr. Friederich Franz, who taught physics at the institute, influenced his later life very strongly.

Johann enrolled at the institute in the fall term of 1840 and tried to find pupils to tutor, as he had at Troppau, but without success. He managed to complete the fall term and part of the winter term before once again becoming so exhausted and ill that he had to leave the institute and go home. He spent most of a year at home under his mother's care getting his strength back, returning to Olmütz in the fall of 1841.

Two things happened during his time at home that helped to make that return possible. The first of these was the sale of the family farm to Alois Sturm, the husband of Anton's older daughter, Veronica. When Anton was injured in 1838, Alois had taken over the running of the farm. Since it was evident that Anton would never be able to work the farm again, and since Johann had decided not to take it over from his father, the next best thing to do was to sell it to Alois, who was at least connected to the family by marriage. The contract between Anton and Alois, which was signed on August 7, 1841, provided, among other things, that Alois should (1) pay Anton 10 florins a year toward Johann's school expenses; (2) set aside the sum of 100 florins to be paid to Johann, "if the latter as he now designs should enter the priesthood"

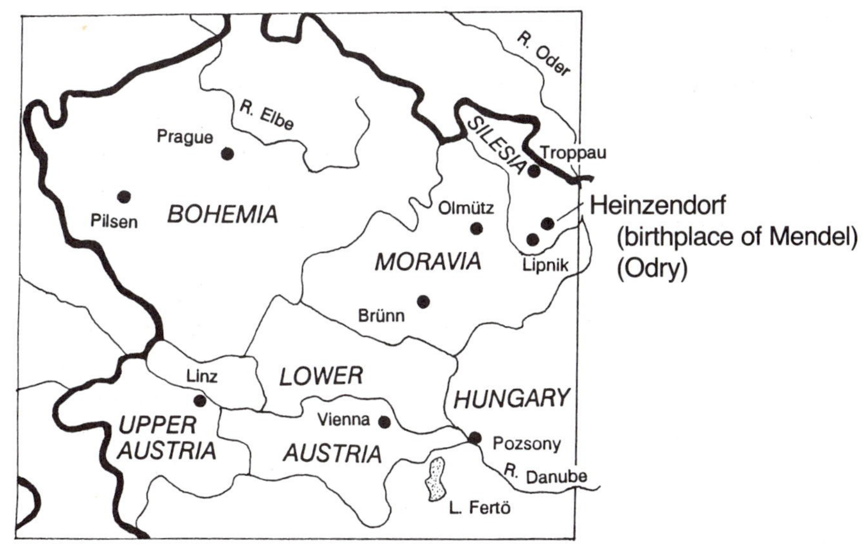

Mendel's Corner of the Hapsburg Empire

(Iltis [1932] 1966, 39); and (3) agree to pay all the costs connected with Johann's celebration of his first mass. From this it is evident that Anton had accepted the idea Johann was to be a priest, as his wife had wished, and not a farmer as he had wished.

There was another clause in the contract of interest to us. A dowry was provided for Anton's younger daughter, Theresia, to improve her prospects for marriage. Partly from love of Johann and a desire to help him finish his education and attain his desired goal, and probably partly with the encouragement of her mother, Theresia gave Johann a portion of her dowry. With this generous help and a better income from tutoring, Johann was again able to take up his studies at the institute.

As is usually the case in educational institutions, not all of the teachers at the Philosophical Institute were equally good. However, Johann was very fortunate in having Dr. Friederich Franz, one of the good ones, as his physics teacher. Father Franz—he was both a priest and an Augustinian monk—had taught physics for twenty years at a similar philosophical institute at Brünn before coming to the University of Olmütz. While at Brünn he had lived at the Augustinian monastery of St. Thomas and still maintained friendly contact with some members of the community. This proved to be fortunate for Johann. At Olmütz

Father Franz taught physics both in the university and in the institute. He replaced Dr. A. von Baumgartner, who had moved to the University of Vienna, where Johann encountered him some years later when he went to Vienna to take an examination for a teaching license. But that is another story. Later, Franz moved on to Salzburg, Mozart's hometown, where he became head of the Modern School. Still later, he became abbot of the Augustinian monastery of Neureisch, a pattern similar to that followed much later by Mendel.

When Johann came to the study of physics with Dr. Franz, he probably received the first clear demonstration of how human intelligence had penetrated the obervable world to disclose the fundamental laws governing the universe. In physics he would have seen how the discovery of constant relationships, through relatively simple experiments, could reveal the natural laws that were the source of these invariant relationships. And the road to such discoveries was paved with mathematics. These discoveries must have had a profound effect on the mind of Johann; physics and mathematics became his favorite subjects, he spent most of his professional life teaching them. They also provided the basic patterns upon which his experiments with peas were based. He must have found the vision of an orderly, predictable world bound together by universal laws to be both sharply in contrast to his own chance-filled struggles and very seductive and pleasant to contemplate. But Johann was not the only one to benefit from this association.

Having Johann as a student must have been a source of great delight and deep satisfaction to Dr. Franz. There are few things more moving and more stimulating to a dedicated teacher than the experience of working with an exceptionally able student who shares the teacher's love of his or her chosen subject. Knowing well the hardships Johann had transcended in order to reach that level in his education could only have increased Franz's respect for him. In such circumstances it is not at all surprising that a solid friendship should develop between him and his student. It must have been a deeply gratifying experience to Johann to have his worth, both as a person and a scholar, recognized in this way. A teacher and a student who have shared such an experience remember it for a long time, and frequently a bond is formed between them that lasts a lifetime. From the little evidence we have, it seems probable that such a bond developed between Friederich Franz, physicist, teacher, Augustinian monk, late of the monastery of St. Thomas at Brünn, and Johann Mendel, peasant's son and dedicated scholar. We shall shortly see what resulted from this.

Early in the summer of 1843, at the age of twenty-one, Johann finished his studies at the institute, graduating again with high honors. The studies there whetted his already keen appetite for further intellectual nourishment. However, his experiences in reaching that point convinced him that he could not even try to go on to a university. He had discovered that there were limits beyond which even his willpower could not drive his body, especially when his utmost efforts provided no more than poor shelter, poor clothing, and totally inadequate nourishment. Unless he could find some means to provide for these needs in a better fashion than he had thus far, further education was beyond his reach. It must have been a bleak and desperate experience to meet this final barrier when he had struggled so hard and his parents and his sister had sacrificed so much in order for him to reach this point.

Inevitably, Father Franz would have been aware of this. Probably, like Father Schreiber and Thomas Makitta, and like Johann's mother far earlier, he felt that it would be an unforgivable waste of a special talent and a fine mind if Johann's intellectual development were to end before he had realized his full potential. Fortunately, just at this point, Father Franz received a letter from one of his friends at the monastery of St. Thomas in Brünn, "requesting him—at the suggestion of the Abbot—to look for possible candidates who might be found to [fill out] the ranks of the monks who had become few in number" (*Messenger* 1938, 7:7). Father Franz had two candidates in mind, only one of whom he felt he could recommend without reservation. That one was Johann Mendel. He wrote to his friend at Brünn; "During the two-year course in philosophy he has had, almost invariably, the most unexceptionable reports, and he is a young man of very solid character. In my own branch he is almost the best" (Iltis [1932] 1966, 42). He asked his friend to pass this information along to the abbot of the monastery, Cyrill Franz Napp.

We must assume that Johann's friend and mentor, Franz, told him what he had done and probably also sketched for him what sort of life he might expect if he decided to enter the monastery. Johann wrote in his autobiography that "his circumstances decided his vocational choice" (Iltis 1954, 234). Having consulted with his parents and obtained their written consent, he took the required physical examination and was accepted as a novice October 3, 1843. As we noted earlier, he then took Gregor as his religious name, the name by which he has been known ever since.

Novice and Theology Student

The Augustinian monastery of St. Thomas, where Mendel was accepted as a novice, is still in existence. It is located in the oldest part of the modern city of Brno, but still within walking distance of the center of the city. In 1583 the margraves, princes of Moravia, had established a nunnery on the site. This was taken over by the Augustinians in 1783 and became the monastery of St. Thomas. At the time Mendel was admitted to the order, Brünn was one of the most important centers of commerce in the Hapsburg Empire and had many beautiful and impressive buildings. It was also the home of scientific and literary societies as well as institutions that served the poor and the handicapped.

The community Mendel joined was an unusual one. First of all, it was not a closed community whose members were cut off from contact with the outside world and spent their time in fulfilling their religious duties and doing whatever work was needed to maintain the community. Instead, the monks were, within limits, free to carry out duties in the outside world, often as teachers in the schools, institutes, and universities of the district such as those Mendel had attended. This points to the second way in which it was unusual. The monastery was not only a center of religious life, but was also a center of intellectual life. Many members of the community were distinguished philosophers, scientists, and artists. Most of them were deeply involved in the intellectual and cultural life of Moravia. Several of them, including the abbot, were very interested in scientific agriculture and were active in the various agricultural societies of the district. Among them was Father Aurelius Thaler, who had maintained a huge herbarium of the flowering plants of Moravia and a small botanical garden. Father Thaler died in 1843, just a few months before Johann entered the monastery. The botanical collection, the botanical garden, and a mineralogical collection that the monastery also owned were soon placed in Mendel's care. As a novice, he spent most of the time he could spare from his required classes studying these collections and working in the garden.

After completing his studies for the priesthood and being ordained, Mendel participated in many of the same activities as the other priests.

Mendel's garden and the entrance to Mendelianum, September 1991 (photo by Mrs. A. F. Corcos)

It is clear that Father Franz's suggestion that Johann should apply to the Augustinians for admission was a good one, for the young man found that the monastery provided a stimulating environment. Here he found shelter and nourishment in plenty for his body, his mind, and his soul. He found himself a brother among brothers, accepted and valued for his character and intellect, despite his family's lack of wealth and titles.

When Mendel joined the Augustinians, the man responsible for shaping the life of the community was Father Cyrill Franz Napp, a very remarkable person. Before becoming the abbot, Napp had been a theologian and professor of Oriental languages at the Brünn Theological College. As abbot he proved to be an excellent administrator, taking care that the physical, spiritual, and intellectual welfare of the community was maintained and enhanced. His "recruiting" of Mendel was part of his continuous efforts to attract individuals of outstanding ability to the monastery and then, by vigorous defense of their rights to carry on independent activities in science and the arts, to keep them and promote their growth. Although these activities frequently brought him into conflict with his bishop, who thought the community was too concerned with such secular activities at the expense of their religious duties, Abbot Napp never failed to support and defend his brothers. As we shall see later, when Gregor Mendel became abbot and came into conflict with the authorities about taxation of the revenues of the monastery, he exhibited the same toughness in defense of the right of the community as had his predecessor. Unfortunately, he was not as successful as Napp had been.

Abbot Napp had many interests and, by reason of his position, wore many hats. He was a member of the provincial Diet, or legislature, where there was little power but much talk. He was also director of higher education for Moravia and Silesia, a position in which he was able to influence strongly the lives of many other students. As we shall see, because of his position, he was able to influence Mendel's future development greatly. He held important positions in the local agricultural societies, a pattern also followed by Mendel when he became abbot. Napp was interested in the improvement of crop plants, fruits, and farm animals and even found time to do a few experiments in the monastery gardens. From this it is easy to see why, at a later time, he was such a strong supporter of Mendel's experimental work with peas. All in all we can see why the monastery of St. Thomas was such a perfect place for Mendel to live, grow stronger in body and mind, and later to come into full flowering with his experiments on hybrid peas. The results of these experiments led, sixteen years after his death, to the creation of a new science, the science of **heredity**. But these developments were still far in the future, for when we left Mendel he was only a novice and not yet ordained.

In preparation for his ordination, Mendel had to complete a four-year series of special courses at the Brünn Theological College. These courses, which he started to take in 1845, covered such topics as church history and law, biblical archaeology, theology and the duties of a pastor. He supplemented his required Hebrew and Greek language courses with optional courses in Chaldean, Syriac, and Arabic. It is highly possible that he was influenced in this by Abbot Napp's reputation as a scholar of Oriental languages. He also attended Professor Franz Diebl's lectures on agricultural economy. In each of these very intensive courses he passed his examinations, as always, with "greatest distinction."

Looking at his school records from his earliest to these latest, there are several things that are very striking. One of these is the wide-ranging character of his education. There had been no narrow concentration in any one area at the expense of everything else, even in his theological training. Second, with the exception of the courses in mathematics and physics at Olmütz, he had little formal training in the physical sciences. A third point to note is his thirst for knowledge: whatever the subject might be, he sucked it up as dry earth soaks up a needed rain. And a last point is the uniformly excellent quality of his learning as judged by his teachers. He was always at or very near the top of his class, not just in one subject, but in all of them. He not only had the necessary intellectual ability but he responded to each opportunity, each challenge, as if driven by an inner compulsion to be at the very top of his group. As we watch Mendel's development from novice to teacher of physics and mathematics, to superb research scientist and pioneer meteorologist, and finally to an abbot of the same high caliber as Father Napp, we need to keep these points in mind. His future development was foreshadowed in large measure in his development up to this point.

Mendel completed most of his preordination studies by 1847, when he was twenty-five. In that year, because some members of the monastic community died unexpectedly, there were not enough ordained priests in residence at the monastery to perform the required daily services. In order to help with this problem, Abbot Napp obtained permission to ordain Mendel a year early on the condition that he must finish his studies the next year. The permission was granted, and Mendel was ordained August 6, 1847.

When Mendel was ordained, he assumed a share of the work of maintaining the daily round of services in the monastery church. Sometime in 1847 he must have celebrated his first mass, but we have no record of what happened to the 100 florins that his brother-in-law was supposed to pay him. The following year he completed the fourth year of his studies and graduated on June 30, 1848, at the age of twenty-six.

One of the most politically turbulent years of the nineteenth century, 1848 was the time when all over Europe voices were raised demanding freedom of discussion, freedom from censorship, and the establishment of liberal constitutions with universal suffrage. The winds of revolution buffeting autocratic regimes had started to blow from the French capital and finally reached Vienna on March 13. They blew away the regime of Metternich, state chancellor for the Hapsburg Empire and a staunch conservative. Even the monasteries were not immune from political turbulence. Mendel and five of his friends signed a petition for more freedom for the members of the Augustinian monasteries, asking in particular that they be free to teach and to do research in science. Except for this little bit of evidence, we know little or nothing of how the revolution of 1848 affected Mendel or his fellow monks. However, that simple fact is enough to suggest that, although Mendel may have been conservative on the surface, he was a reformer at a deeper inner level. After all, he had another reason besides his concern about the freedom to teach for following events closely, even if at a distance, since one of the demands put forward by the would-be reformers was the abolition of the feudal labor dues, the robot, which had made life so difficult for his parents and himself. It was in fact abolished forever in 1848.

From July 1848 to March 1849, as the result of a double compromise, the Hapsburg Empire was a constitutional monarchy: the reformers accepted the empire, and the emperor tolerated reformers. However, the compromise did not last, and the government resorted to armed force to restore law and order. A new wave of reaction, absolutism, and repression engulfed the empire for another ten years.

Having completed his studies and been ordained, Mendel was given several new duties in addition to his work in the monastery. He was assigned as an assistant pastor at a church in Brünn, and Napp assigned him to serve also as a chaplain at a hospital near the monastery. Many of Mendel's new parishioners and many of those in the hospital spoke only Czech. In order to talk with these folk and to preach to them in church, he had to improve his knowledge of that language. In addition, he was studying various other subjects, especially the natural sciences, preparing to take the examination for the doctor of philosophy degree. However, for a combination of reasons, he never took the examination and so never became Dr. Mendel.

At the time he was supposed to take the examination, he was not in Vienna, where it was being given, but was serving as a temporary teacher in the gymansium at Znaim (now Znojmo). This abrupt change in his plans and role came about because he found that he could not carry out some of the duties required of him as a parish priest and hospital chaplain. He could not bear to be in close contact with people who were ill or in pain. As a pastor he was expected to visit those of his parishioners who were confined at home because of illness or injury, bringing them the sacraments, joining them in prayer, and offering them spiritual guidance and comfort. When the services and prayers were finished he became very uneasy and almost unable to speak comforting and helpful words to them. Visiting the wards in the hospital was even worse. Making his rounds of the parish and the hospital wards became an ordeal for Mendel, and he soon became so seriously ill himself that Napp had to relieve him of these responsibilities.

Fortunately, Abbot Napp was in a position to offer Mendel another occupation shared by many Augustinians, teaching in the schools of the empire. We have already seen them in these roles at Troppau and at Olmütz. These were only two local examples of the general importance of the Augustinians in the schools, institutes, and universities of the empire. At this time, in the mid 1800s, they filled the same kinds of roles that the Jesuits had filled in the eighteenth century. Given these kinds of expectations for Augustinians and his position as abbot of an important monastery, it is not surprising that Napp was director of higher school education in Silesia and Moravia. Thus, when a vacancy occurred at the gymnasium at Znaim, the request for a temporary teacher to fill the vacancy came to Napp. The request was for a substitute to teach a new seventh class just being started there. In response, Napp sent Mendel's name forward as a suitable candidate. At that point Mendel was probably ready to agree to almost any change if only it took him away from having to face more sick people.

The governor of the district agreed to Napp's proposal, sent Mendel formal notice of his acceptance, and ordered him to go to Znaim forthwith (Znaim was some thirty to forty miles southwest of Brünn). The original appointment turned out to be wrong in regard to what he was to teach on nearly every point, but Mendel adapted to the changes and started teaching on October 7, 1849. Although he was not at Znaim for very long, he won the goodwill of his pupils and, when he left after a few months, received the most glowing testimonials from his fellow teachers and superiors. They praised the clarity of his explanations, his knowledge of his subjects, his diligence, and the sound religious, moral, and ethical values displayed in his teaching. They considered him to be "an exemplary and safe teacher of youth." He had not uttered so much as a single "repulsive word about religious principles or political or regulations" (*Messenger* 1938, 10:11).

If these testimonials about Mendel's character and political safety seem to us overblown and obsequious, we need to remember that they were written in 1849 in the Hapsburg Empire, which was then a police state. This was only one year after the revolutionary year of 1848, when for a while everything that had been nailed down for generations was coming loose. Everyone was uneasy, and, after all, Mendel had signed that petition. In such circumstances, the testimonials would serve to demonstrate the political orthodoxy of everyone concerned: Mendel, the administration of the gymnasium, his fellow teachers, and, presumably, Abbot Napp as well. This was only the first of many such testimonials Mendel was to receive in the course of his life as a teacher and later as an administrator.

Since a regular teacher was soon obtained, Mendel returned to the monastery, but for only a short while. Even before the testimonials from Znaim reached him and Napp, he took up a new assignment that was more in his line and also in Brünn. The professor of natural science at the Brünn Technical Institute, Dr. Helcelet, became ill and a temporary replacement was needed. Again Mendel was called in as a substitute teacher. Again he served for only a short time—April 7 to June 6—at which time Dr. Helcelet was well enough to return to his duties. In this position, as at Znaim, Mendel's personality and talent again won him affectionate praise. It was evident to the teachers and students with whom he worked at Znaim and at Brünn that Mendel had a true vocation as a teacher, especially as a teacher of science.

However, if he was to continue in that vocation, he needed to obtain a teacher's license from the government. Thinking highly of him, the faculty of the gymanasium at Znaim decided to propose him as a candidate for the examination that would license him to teach natural history at all levels of the curriculum and to teach physics in the lower school. The application was filed on April 15, 1850. Mendel wrote a brief autobiography to accompany it, which was dated April 17, 1850, and offers us one of the few glimpses of the way he saw himself at age twenty-eight (Iltis 1954). Applicants were usually accepted only after finishing a number of years of university study and, as we know, Mendel had no such credentials. Therefore, it is surprising that he was accepted as a candidate. Because he was no longer teaching, he had more free time in which to prepare to take the examination for the license.

The examination consisted of two parts, one written with reference books available to the writer. The second part consisted of two sections, one written without references and the other an oral examination. Mendel received the first set of two questions, one in physics and one in geology, in May of 1850. He was given six to eight weeks to write out his answers and send them to the University of Vienna for evaluation. Once the examiners had reviewed the first set of questions, they would decide whether the candidate should report to Vienna for the second part of the examination. Mendel's essay on the physics question satisfied the examiner, Dr. A. von Baumgartner, completely. However, the examiner in geology, Dr. Kner, was very dissatisfied with Mendel's essay in that subject. Mendel's fate was left hanging in the balance pending his performance on the second part of the examination.

Mendel's results on the written section of the second part were much the same as the first: limited success on the physics question and failure on the natural history question. Even so, he was allowed to appear for the oral section of the examination. Both examiners found his results on this to be unsatisfactory and gave him a year in which to improve his preparation. He might then sit for the examination again. Fortunately, his situation was not as dark and discouraging as it may have appeared at the time. When Abbot Napp wrote to Dr. Baumgartner to ask what had gone wrong, why Mendel had failed, Dr. Baumgartner wrote that he was impressed by how much Mendel had achieved, almost entirely by himself, and that he felt that Mendel had real talent and deserved assistance with some formal training at the University of Vienna. As a result, Napp arranged with the bishop for Mendel to spend the next two years in such a course of study.

Mendel left Brünn on the night train to Vienna on October 27, 1851. Arriving in Vienna, he enrolled at the university in the philosophical faculty as an extraordinary student, or, as we would put it, as a special student. His classification as an extraordinary student meant that he did not have to take the prescribed (or required) curriculum of that faculty but was free to select courses that would serve his special needs and interests. We must suppose he had some guidance in making these choices to ensure that he did not sign up for courses in which his background was too poor for him to have any chance of success.

Thus, in the fall of 1851 at the age of twenty-nine he embarked on a series of experiences that were to convert him from a gifted amateur, deeply devoted to the study of natural sciences, into a very well trained experimental scientist of great originality and power. Once again, generous and perceptive friends had conspired to convert what seemed a dead end into an open door to a whole new world of ideas. Once again, someone had felt that he had too good a mind to waste, too excellent a talent to allow it to be lost. Without these generous gestures and

Mendel's eager response he would not have carried out the remarkable series of experiments that resulted in this great text in science, *Experiments on Plant Hybrids*.

University Student

The Vienna to which Mendel came in the fall of 1851 had recovered from the worst excesses of the 1848 revolution. The foreign radicals and rabble-rousers had been driven out, and the university students had, for the most part, returned to their studies. The new emperor, Francis Joseph, had come back to the city escorted by loyal units of the army. The musical and legitimate theaters had reopened. Composers, performers, and artists from the empire and the rest of Europe had recreated the musical life for which Vienna had long been noted. The rich cultural life of the city had been restored.

Undoubtedly, there were feelings of bitterness that hopes for greater freedom had been frustrated, feelings of anger at injuries to persons and property. And, although the university students had settled down, at least for the moment, they were still regarded with hostility and suspicion. Thus, when Mendel arrived to become a university student he carried with him his passport, as required by law. That he was an ordained priest and a monk did not exempt him from this requirement. From this document, which still exists, we know that he was then twenty-nine years of age, of medium height, with light hair and grey eyes. He had no distinguishing marks, and his native language was German. From the reminiscences of his students and some photographs we know that his hair was curly or wavy. The document also contains a record of each of his departures from Vienna and his returns during the two years that he was a student at the university.

Having found approved lodgings near the convent of St. Elizabeth, Mendel proceeded to enroll in the university. He intended to take courses there that would fill in the gaps in his education revealed by his failure on the examination for a teaching license. However, serendipity took over, and the man who went to Vienna to become a better teacher of physics and natural history acquired from his teachers the techniques of a scientific researcher. He caught along the way their vision of the natural world as an orderly world, governed by natural laws waiting to be discovered. They also taught him that the search for those laws was the great adventure, their discovery the highest reward a scientist could receive.

The science faculty of the University of Vienna with whom Mendel studied included some of the most distinguished scientists of that time. One of these men was the physicist Christian Doppler, after whom a famous physical law is named. Doppler discovered this law while studying a phenomenon most of us experience every day when we are standing on the curb and a rapidly moving car approaches us with the horn blowing. As the car comes toward us and passes by, the pitch of the horn *appears* to change. The actual pitch of the horn has not changed; it only appears to do so because of the motion of the car. The reasons for the apparent change were first studied quantitatively by Dr. Doppler. Since he was the discoverer of the relationship, expressed in a mathematical equation, it became known as Doppler's law and the apparent change in pitch as the Doppler effect. This same relationship is used to study the relative motions of the Earth and distant stars and galaxies, studies that have resulted in the belief that the universe is expanding. This is an example of the high caliber of the men with whom Mendel was to study for these two vital years.

When Mendel enrolled at the University of Vienna, Dr. Doppler was nearing the end of his academic career but was still teaching courses in experimental physics and instructing students in preparing and doing experiments demonstrating important laws and relationships in physics. Mendel signed up for Dr. Doppler's lectures the first semester he was at the university and continued to study at his institute for the next two semesters. There is some evidence suggesting that he was very successful at the institute, serving for some time as an assistant demonstrator. This was something of an honor in the circumstances, since he had not yet completed his course work. We know that afterward, when he returned to Brünn and became a teacher of physics, he had a reputation for doing excellent demonstrations. From his own words, written to the Swiss botanist Karl von Nägeli, we know that he thought of himself professionally as a teacher of experimental physics rather than as a botanist or hybridizer.

In the spring semester of his second year at the university he took another course on making and using physical apparatus and also one on higher mathematical physics. These courses, together with some in pure mathematics, add up to more than half of the seventy-two credits he took in the two years. When we add to these several courses in chemistry, the total is about 70 percent of the credits he took. Thus, the courses he had chosen in these areas served very well to fill the gaps noted by Dr. Baumgartner and Dr. Kner when he failed on his first trial for a teaching license. When he had completed this part of his studies, he was fully up-to-date on the major concepts and systems of ideas of physics and chemistry, both theoretical and practical. He was also well grounded in the relationships of experimental methods to the search for the laws of nature. As we shall see when we study his experimental work with pea hybrids, these experiences and this training played a major role in what he decided to look for and in the way he designed and interpreted his experiments. Without the education he received at the University of Vienna he could not have conceived the problem he worked on as he did. Without this training there would have been no *Experiments on Plant Hybrids*.

But what about the other 30 percent of the courses he took at the university? What did he take and what did they contribute to his future work? These courses included, among others, the description and classification of plants, the physiology and paleontology of plants, systematic zoology, and exercises in the use of the microscope. The relationship of the courses in zoology and paleontology to his problems with animal classification and in geology on the licensing examination are clear enough. In fact, some of these courses were taught by Dr. Kner, the man who had failed him on these portions of the examination. The courses in botany reflected Mendel's interest in plants going back to his childhood experiences. Among the botany faculty, two men stand out because of their influence on Mendel. These men were Edward Fenzl and Franz Unger.

At the time when Mendel came to the university the Botanical Institute consisted of two departments; the Department of Plant Anatomy, headed by Professor Fenzl, and the Department of Plant Physiology, headed by Professor Unger, the first professor of this subject in the Hapsburg Empire. The courses Mendel took with Fenzl provided him with a solid grounding in the classification and description of plants but did not inspire him or inflame his imagination. Dr. Fenzl was essentially conservative in his scientific views, concentrating on transmitting to his students the best knowledge of his subject matter then available. He still held to the old view that there was a force that propelled and guided the development of all living things. This force, historically known as *vital force*, or *vis viva*, was different in nature from those Mendel studied in physics and chemistry. Dr. Fenzl's views would have pointed his students to the concept that species of plants were essentially stable and that many if not most of them had existed virtually unchanged since the beginning of time.

The teaching and personality of Professor Unger differed in almost every respect from those of Professor Fenzl. Where Fenzl was conservative in attitude, Unger was essentially progressive, emphasizing the latest discoveries in the anatomy and physiology of plants. He turned his back on the past and regarded the development of living things as being directed by the same physical and chemical laws that governed changes in the inorganic world. He emphasized the importance of studying variation, especially the process of reproduction, in plants as a key to understanding the ways in which contemporary plants had changed from earlier forms found as fossils in rocks. He repeatedly called his students' attention to how much was still unknown and offered fruitful suggestions for directions in which to investigate. It is easy to see how such a man, with his learning lighted up with imagination, to paraphrase Whitehead, could evoke a strong response from his students and especially from Mendel.

One of Unger's conceptions seems to have lodged firmly in Mendel's mind. Unger stressed the idea that understanding the processes of cellular reproduction and fertilization was critical for understanding variation in plants. He recognized, with J. M. Schleiden and T.A.H. Schwann, that organisms are communities of cells in which each cell is an independent unit of structure, new cells developing from old. Unger also believed that each new plant resulted from the union of one pollen cell and one egg cell, an idea not generally accepted at that time. In addition, the laws governing the development of plant characteristics as generation followed generation were then totally unknown. The idea that there should be definite laws governing the development of plant characters took root in Mendel's mind, and shortly after he returned to Brünn at the end of his university studies he embarked on his series of experiments with pea hybrids. The whole purpose of these experiments was to discover the laws guiding the development of observable characteristics in hybrid peas generation after generation. The torch had been passed from Unger to Mendel.

Mendel is sometimes thought of as an amateur botanist. If the word *amateur* is understood to mean that he was not a professor of botany in a distinguished university or a member of an institute, then Mendel was an amateur, but so were many other famous scientists, including Darwin and Einstein. On the other hand, if *amateur* means that he had not received any formal training in the field, Mendel was not an amateur at all, since he was as well trained as anyone whose destiny was to be a professional researcher.

Teacher

At the end of summer 1853, having finished his studies in Vienna, Mendel returned to the monastery of St. Thomas. We do not know what else he may have done between then and May 1854, but we do know that in 1854 he turned to teaching again because at that time he was appointed teacher at the Brünn Modern School. This school had been founded recently to train young men for industrial jobs. There Mendel taught physics and natural history in the lower classes, which were divided into sections of sixty students. His immediate superior was Alexander Zawadski, who earlier had been professor of applied mathematics and physics at the University of Lwów. Because Zawadski had taken part in the political movement of the revolutionary years of 1848–1850, he had been downgraded to a high school teacher. However, his scientific interests remained as numerous as before. Besides mathematics and physics, they included botany, zoology, paleontology, and evolution.

Mendel stayed at the Modern School for the next fourteen years, teaching physics and natural history. Here also he was an outstanding and inspiring teacher, extremely well liked by his students and respected by his colleagues. Desiring a permanent teaching certificate, he applied a second time to the examining board in Vienna in 1856 and was accepted for examination. However, he was never certified. It has been suggested that he became ill during the examination and never completed it. We know that at times during his student years he experienced some kind of stress-related illness on several occasions. Certainly, facing this examination must have been a very stressful situation. Given his long record of academic achievement and his great success as a teacher, it is probable that everyone who knew him expected that he would come through splendidly. Strangely enough, unlike his first examination, where we have a very full record, here we have no answers written out, no record of any kind to help us understand what happened. All we know is that he made no further attempt to obtain a state teaching license and remained an instructor at the Modern School in Brünn, where he did not need any certificate to teach.

Mendel's Peas

For the remaining years of his life, Mendel lived at the monastery of St. Thomas, a peaceful and quiet place surrounded by large gardens and a small wooded area where one could easily meditate and study. The monastery and this land were enclosed within high walls. It was on these grounds, below the windows of his room, that Mendel had his experimental garden. There, in 1856, he started his pea hybridization experiments, the results of which were eventually to immortalize his name. For eight consecutive years he grew thousands of peas on a 35 × 7 meter plot (115 × 23 feet). Mendel finished his experiments in 1863 and in 1865 presented his findings in two evening lectures before the Natural Science Society of Brünn, which he and some of his friends and colleagues had founded in 1861. In 1865 the society membership had reached 162 and included representatives of a variety of disciplines—geology, botany, physics, astronomy, and so on.

According to the local newspaper, the *Brünn Tagesbote,* there was "lively participation by the audience," but we do not know what the nature of this participation was. Nor do we know the reaction of Mendel's immediate friends to the results of his experiments. It is surprising that there is no written record of what they thought since many of them shared his interest in hybridization. However, they may not have shared Mendel's enthusiasm for his pea experiments because they were more interested in learning about hybridization of wild species and its relation to evolution than about hybridization of vegetables. Their interest in evolution can be understood in the light that Darwin's *Origin of Species* had been translated in German just three years before. The botanists among them were not interested in finding the laws of hybridization as Mendel was but were interested in whether or not stable, **true-breeding** hybrids could be found in nature.

It is even more surprising that we do not know what Abbot Napp thought of Mendel's pea experiments. After all, Napp's enthusiasm for research in breeding and heredity was well known, and he had morally and financially encouraged and supported Mendel in his research.

One year later, in 1866, Mendel's lectures were published in the

Proceedings of the Natural Science Society of Brünn (Verhandlungen des naturforschenden Vereins in Brünn) under the title "Experiments on Plant Hybrids." Hoping to hear from the scientific community, he obtained forty reprints of his published lectures to send to prominent botanists and researchers in hybridization in the major western countries. Only one answered and then only after Mendel wrote him a personal letter. It is known that some who received a copy never read his paper, since many years later their reprints were found uncut on the shelves of their libraries. The single exception was Karl von Nägeli, a man deeply interested in hybridization and the man to whom Mendel had written. From December 1866 to November 1873, Mendel sent a series of letters to him concerning his pea (and later, hawkweed) experiments. After Nägeli's death these letters were collected by his family and given to Carl Correns, a former student, who published them in 1905. Of the letters from Nägeli to Mendel, only part of one letter has been recovered.

From Mendel's letters to Nägeli, it seems that Nägeli was far more interested in Mendel's work with hawkweeds than in his work with peas. Although he understood the experiments, he did not seem to understand their significance. He urged Mendel to continue working with hawkweeds, which was unfortunate because hawkweeds, unlike most flowering plants, produce viable seeds most of the time without the flowers needing to be fertilized with pollen, the male contribution. As a result, they produce seeds of only maternal origin. When these seeds are sown, the resulting plants are all like the mother. By contrast, hybrid seed may produce several kinds of offspring. In hawkweeds, hybrid seed is produced only rarely. Not knowing this, Mendel worked for the next five years attempting, mostly without success, to reproduce with hawkweeds the results he had obtained with peas.

Abbot Mendel

In March 1868, after the death of Abbot Napp, Mendel, then forty-six years old, was elected by his peers to become the abbot for life of the monastery of St. Thomas. How this happened is an interesting story. Unlike other Augustinian monasteries, the one at Brünn had been given the right to elect its own abbot. However, the election had to be confirmed by the emperor, who for political reasons preferred the abbot to be ethnically German. Most of the monks at St. Thomas at that time were Czechs, so the choice was restricted to two monks, Mendel being one. After five ballots spread over two days, Mendel, who was politically acceptable and also liked by his fellow monks, was unanimously elected.

The choice was a fortunate one. Abbot Mendel won the respect of all, both as a person and as an administrator. Like his predecessor, Napp, Mendel kept the monastery in excellent financial shape. He became well recognized for his administrative abilities, not only as head of the monastery, but in other capacities such as the acting head of the Agricultural Society, the curator of the Institute for the Deaf and the Dumb, which was situated in Brünn, and as an officer of the Moravian Mortgage Bank, which he served first as deputy manager and later as manager. His performance as an administrator was remarkable, for when Mendel was elected abbot he had no administrative experience whatsoever, having spent most of his adult life as a science teacher.

However, he appears to have been a very careful observer of the practices of Napp and to have come to the conclusion that some of them needed to be changed. According to the Reverend A. G. de Romanis (1929): "Under the direction of his predecessor, Abbot Napp, there was not much zeal in the observance of the rules, and community life was rather *sui generis*. Mendel allowed no deviations from the rules of the Order and made short work of the excuses and falterings." Under Mendel the buildings of the monastery were restored and redecorated, especially the abbey church, which also served as the parish church for the district. In addition, he restored the services of the church to their full and proper forms. Like Napp, he recruited to the order the best men he could find and then did whatever was necessary to promote their growth and development both intellectually and spiritually.

Death of Mendel

The last years of Mendel's life were saddened by his long dispute with the Hapsburg government about taxation of the monastery. In 1874, the Imperial Government passed a law decreeing that each of the monasteries of the empire had to contribute a certain amount of money each year to the Ministry for Public Worship and Education. Abbot Mendel considered this law a gross injustice and refused from its beginning until his death to make the payments imposed on the monastery. Orel believes that Mendel's health was seriously weakened by this long struggle. Feeling the stress, Mendel wrote to the provincial governor that in the last two years he had "prematurely aged" (Orel 1984b, 231).

During the last few years of his life Mendel developed Bright's disease. He had increasing difficulty getting around because of the accompanying heart trouble and dropsy. The final causes of death were kidney failure and congestive heart failure. He died on January 6, 1884. On January 9 he was given a magnificent funeral, which was reported in the local evening newspaper, the *Brünn Tagesbote* (Matalova 1984, 220).

Mendel's activities in the various fields of banking, agriculture, meteorology, and teaching were recognized in many obituaries that appeared at his death in the local newspapers and in the journals of the societies to which he belonged. His constant interest in the practical affairs of Moravia was well recognized. Notice also was taken of his kindness toward the poor, whom he had helped on many occasions. Nor was his work with peas forgotten, at least not by the **pomology** section of the Agricultural Society. In its orbituary of Mendel the last sentence reads: "In fact his experiments with plant hybrids opened a new epoch—What he did will never be forgotten" (Matalova 1984, 219).

As a matter of fact, however, no one paid much attention to Mendel's paper until 1900 when, within the space of a few months, it was resurrected by three biologists: Hugo de Vries, professor of botany at the University of Amsterdam, Holland; Carl Correns, a botanist at the University of Tübingen, Germany; and Eric Von Tschermak, an assistant at the Agricultural Experiment Station at Esslingen, Austria. Though their interpretation of Mendel's paper is at variance with what Mendel tried to convey in 1866, they are responsible in great part for establishing him as one of the greatest scientists.

Reflections on Mendel, the Scientist

One year before his death, Mendel reportedly said to a new priest making his monastic vows: "My scientific work has brought me a great deal of satisfaction and I am convinced that it will be appreciated before long by the whole world" (Orel 1984a, 99). By 1900, Mendel's prophecy of 1883 had been fullfilled.

What Mendel did not tell us is what scientific work he had in mind when he said this. People have, for a long time, assumed that he was thinking about his hybridization experiments with peas when he made the above statement. But these experiments were only a part of his scientific work. In fact, Mendel was interested and active in two very different fields of science: meteorology and breeding of plants and animals.

For much of his life Mendel was fond of breeding new varieties of ornamental flowers, vegetables, and fruit trees. Though he was extremely busy after becoming an abbot, he never lost his interest in hybridization and he had plantations of strawberries and black currants on the grounds of the monastery. From 1864 to 1873 he performed an extensive series of experiments with many different kinds of plants. From his correspondence with Nägeli we know that Mendel had much trouble with many of these experiments and finished only a few. He was able to confirm the results he obtained with peas with some kinds of plants but not with others.

Mendel not only bred plants, but hybridized bees. Interested in new races of bees better adapted to the region, he built a beehouse in back of the monastery in which he kept both local and foreign varieties. Despite extreme care in his breeding methods and keeping the beehouse in perfect condition, he was unsuccessful in producing new varieties because, as we know now, reproduction in bees is highly complex. His fondness for his bees was well known. He demonstrated such a love by having a denuded hilly area on the south side of the monastery planted with all kinds of flowers so his bees would have a large source of honey. When Mendel decided to redecorate the Great Chapter Hall of the monastery in 1875, he had a beehive painted on the ceiling. Mendel seemed to have a fondness for any organism he worked with, and he used to refer to his peas as "my children" and frequently talked about his bees as "my dearest little animals"(Matalova and Kabelka 1982, 209)

The second field of science in which Mendel was deeply interested was meteorology, to which he contributed much time during many years of his life. It seems that his interest in this science was born when he was a student at the gymnasium of Opava. His interest can be explained at least in part by his background. As the son of a farmer he was completely aware of the tremendous influence of the weather on the productivity and harvesting of crops. By making precise meteorological observations he might learn how to make reliable weather predictions which would be of great value to farmers. His contributions to the field started in 1848 when he made his first official meteorological observations. From 1848 to 1862 he and Dr. Pavel Olexik recorded a series of observations and they presented them at the first meeting of the newly established Brünn Natural Science Society in 1861.

Later Mendel supplied the Vienna Meteorological Institute with the data of his daily observations in Brünn. The institute had begun publishing daily weather reports and forecasts in 1877 after experts such as Mendel had testified to their value for farmers, especially at harvest time. In concert with his meteorological studies, Mendel kept records of the levels of ground water and also made a long series of observations on sunspots as well as measurements of the amount of ozone in the atmosphere. His interest and experience in meteorological phenomena permitted him to describe in great detail the results of a tornado that struck the town of Brünn and the monastery on October 13, 1870. In the course of this, he observed the speed with which the funnel was moving and rotating, and proposed a theory for its formation. He gave a thoroughly scientific account of the tornado, free of any religious overtones, at a monthly meeting of the Brünn Natural Science Society. The account, which was published in the next volume of the *Proceedings* of that society, is very important because it also shows how deeply the scientific view of nature was ingrained in Mendel. Note that his meteorological work was well known and highly regarded during his lifetime, but is now almost forgotten. By contrast, his work with peas, to which he devoted only eight years, was ignored during his lifetime, but is now highly regarded.

As we mentioned earlier, throughout his adult life Mendel was an active member of various agricultural-scientific societies. Among them were the Agricultural Society of Moravia, the Zoological-Botanical Society of Vienna, and the Natural Science Society of Brünn founded in 1861. As early as 1851 he had joined the Agricultural Society of Moravia as a member of the natural science section. In 1863 he was elected a committee member of the pomology-viticulture section. Soon after becoming abbot, Mendel was elected to full membership on the Central Board of the Agricultural Society, just as Abbot Napp had been. Election to such a board was an honor, but also involved obligations and much responsibility. As a member of the executive board, Mendel devoted great attention to the promotion of agricultural education and agriculture in general. He often had to take care of the current affairs of the society in the absence of the president and vice-president, who were often very busy with their parliamentary duties. Also, during his tenure on the executive board of the Agricultural Society, he was in charge of government subsidies, cooperated in propagating new varieties of crop plants, attended animal husbandry improvement workshops, was in contact with the first farm-machinery cooperatives, supported agricultural exhibitions, granted scholarships, awarded society medals, and contributed to the organization of agricultural libraries. In addition, to all these activities, from 1870 Mendel was the editor of the the Agricultural Society's journal. In 1877, in appreciation of his work, he was asked to be president of the society. He did not accept this honor because he was already ill. Despite his illness, however, he participated in all the monthly meetings.

Reflections on the Human Side of Mendel

If we are to understand the human side of Mendel we must always remember that he was a farmer's son. Although not destined to become a farmer himself, he nonetheless carried the impress of those early formative years with him throughout his life. Jacob Bronowski, scientist and popularizer of science, has written that Mendel remained a farm boy all his life (Bronowski 1973, 115). Mendel's experience as a child spending much time out-of-doors intimately influenced by the weather and by his contacts with growing things, especially plants, left its mark. His experience with the annual cycle of the seasons, each with its established tasks, also influenced him, as did the necessity of placing the performance of the daily round of chores ahead of personal feelings and desires. He must also have learned that no matter how faithful and diligent farmers might be, they were also subject to chance events—natural, political, and social—that were beyond their control.

Superimposed upon this harsh but bracing background was the warmer, softer foreground of a loving family. Mendel shared with his father and with Father Schreiber a love of working with plants. His mother encouraged the growth of his body and of his rapidly developing intelligence. In this last she had the support of Father Schreiber, who shared with the Mendels his love of ideas and the comforts of the church.

Even after entering the monastery at Brünn, Mendel always kept in close contact with the members of his family. In their letters to him, they kept him informed of what was happening on the farm. He wrote them numerous letters in reply, sharing with them their joys and sorrows. He also wrote about important events that affected the country.

For example, in a letter to his mother in 1859, he deplored the bloody effects of war, but at the same time he expressed his patriotic desire that the enemy be defeated.[2]

Mendel always attempted to help his parents. As a priest he could do little financially, and though it became easier for him to help once he became an abbot, by that time his father had died. In a letter to his nephew Ferdinand Schindler, he wrote of his mother: "Even if I had not learned to know the Fourth Commandment, I would ever feel obliged in my heart to help her, to lighten the burden of old age as much as it is in my power, for she has been at all times a good mother" (*Messenger* 1938, 17:18).

Mendel especially showed his generosity toward the children of his younger sister, Theresia, who had helped him to finish his secondary education by giving up part of her dowry. He never forgot her kindness and in return he sent his three nephews, at his own expense, first through high school and then to the university. Two became physicians and one became a teacher of geodesy and astronomy. Mendel and his nephews were very close, sharing among other things a love for playing chess.

We have noted above how he maintained his contact with his family through letters and later through his nephews. When he learned that a number of the villagers of Heinzendorf had lost their homes through fires he contributed a substantial sum of money to help the villagers found a fire department. This was so unusual and so valued that Mendel was given "the freedom of the burg" (Iltis [1932] 1966, 244) and made an honorary member of the fire brigade. When he died in 1884 the brigade sent a delegation to march in the funeral procession.

Though Mendel was a priest, his scientific papers were free of any religious comment. It is therefore of interest to ask the question, how religious was Mendel? Was he a free thinker and only an official Catholic, as his first biographer, Iltis, has suggested? Or was he, as Bateson, one of the first geneticists, has suggested, one of those "numberless honest men in the world who can take things as they find them . . . [men] who are not troubled by questions of faith or doubt, keeping their minds in well divided compartments" (Bateson 1928, 34).

The evidence suggests that Mendel took his religious duties very seriously. When he was elected abbot of the monastery he made many changes in the life of the community. Although he insisted upon the monks living and acting according to St. Augustine's rules, he appeared not to have made life in the monastery grim and harsh. When he had to reprove or correct someone, he did so gently. It was his wish that "his monks should feel that the monastery constituted a large family where they should find pleasant companionship" (*Messenger* 1938, 23:11).

As part of his concern for the religious life of the community he enhanced the beauty and impressiveness of the services, ordering splendid new robes. He also had a new organ built and installed, and supported Father Pavel Krizkowsky's musical activities. Father Krizkowsky, a noted church musician and teacher of music, had among his students Leos Janacek, who became a well known organist, choirmaster, and composer whose music is still played today. Through activities such as these, Mendel strove to maintain and enhance the status of the monastery as a cultural as well as a religious center in the life of Brünn. When he died the music of his requiem was conducted by Krizkowsky's student Janacek.

If his scientific papers show no evidence of religious comments, what do we know about the presence of scientific references in Mendel's religious writings? We have only two pieces of evidence, which can give us at least a little insight. As abbot he delivered sermons on the great feast days of the church year. His notes for two of them have been recovered. They give the impression that Mendel was a down-to-earth priest rather than a man of theological erudition and sophistication. One of them, given during Easter week, illustrates this. Mendel preached on the story from the Gospel of John (chap. 20), which tells how Mary Magdalene saw the risen Christ in a garden and mistook him for a gardener. Mendel commented: "A gardener plants the seed or the seedling in the prepared soil. The soil must exert a physical-chemical influence so that the seed or the plant can grow. Yet it is not sufficient. The warmth and light of the sun must be added together with rain in order that growth may result" (Zumkeller 1971, 249)

A theologian interpreting the above words of Mendel is likely to turn the homely image of a gardener (and the scientist's interest in the physical and chemical action of the soil and the need for sun and rain) into a dissertation on divine grace, which nourishes a person's spiritual life and yet requires human cooperation, just as the gardener cooperates with nature. Simple people may have just remembered the story, the gardener, and the literal statement of the process of growth—so close to the heart of Mendel, the farmer's son who had become a priest and an abbot.

Unfortunately, we do not have enough examples to know whether this sermon was typical or to determine just what kind of spiritual message Mendel actually attempted to convey. But it looks as though Bateson was partly right: Mendel kept scientific ideas separate from religious ideas, but he might not have always kept his sermons free of scientific ideas.

The question of how liberal Mendel was has also been raised. Two events in Mendel's life bear on this question. He signed a petition calling for civil rights in 1848. Certainly, signing the petition, which was written in strong language, indicates that Mendel was a man with great courage and a progressive mind. The second event in Mendel's life occurred two years after the fall of the Bach reactionary government, when in 1861 he and his friends decided to found the Natural Science Society of Brünn. They saw an opportunity to break away from the Agricultural Society, which they thought was in the hands of conservative landowners. They established a society devoted more to pure science than to practical agriculture.

Whether we consider him to have been a liberal or a conservative, our Mendel was an exceptional person. He rose through his intelligence, industry, and tenacity to a position of honor and authority beyond his wildest boyhood dreams. Let us end this painting of Mendel's character with some of his own words, written when he applied to take the examination for a secondary teaching cerificate and had to submit a short sketch of his life. He ended it with these revealing words:

> The respectfully undersigned believes to have rendered with this a short summary of his life's history. His sorrowful youth taught him also to work. Even while he enjoyed the fruits of a secure economic position, the wish remained alive within him to be permitted to earn his living. The respectfully undersigned would consider himself happy if he could conform with the expectations of the praiseworthy Board of Examiners and gain the fulfilment of his wish. He would certainly shun no effort and sacrifice to comply with his duties most punctually. (Iltis 1954, 232)

Notes

1. A *gymnasium* is a secondary school offering an academic classical education and leading to a university matriculation. The type of curriculum offered by the gymnasium is opposed to that offered in a *realschule*, which is characterized by an emphasis on the study of sciences, mathematics, and modern

languages. The realschule rose in Germany in the early part of the eighteenth century, when the tremendous progress of science compelled, after a long struggle, the recognition of this type of school as of equal rank with the gymnasia in the national system of education.

2. The war to which Mendel referred was the one in which the French and Italian coalition was trying to regain for Italy the provinces in Italy held by the Austrians.

Introduction to Plant Breeding

The origins of plant breeding can be traced back to the very origins of cultivated plants. However, no one knows for sure when the domestication of plants occurred. It is thought to have been as early as eleven thousand to sixteen thousand years ago in the highlands of what is now the Middle East. The shift from food gathering to food production did not, however, happen only in one place and all at once. It happened in the Old World as well as in the new. Plant remains were left by lake dwellers in Switzerland, in the barrows of the British Isles, in the ruins of Ancient Mesopotamian and Egyptian cities, and in the caves of Mexico and Peru.

These plant remains indicate that most of the crops we have now were already cultivated three thousand to six thousand years ago and that they were modern looking, far different from their wild ancestors. For example, lima beans found in the ruins of some of the oldest civilizations of Peru have seeds that are nearly one hundred times as large as those of wild limas of today. From an agricultural point of view the most striking and interesting differences between cultivated and ancestral forms of our crops may be found in the grass family, to which cereals belong. For thousands of years our ancestors have collected grass seeds for food. This is still done in some parts of Africa and in North America where wild rice is harvested around the Great Lakes and in Minnesota lakes.

In wild rice, as in other wild grasses, a notable feature is that the **inflorescence** (the flower cluster) can easily break into pieces when the grain is ripe. The breaking up of the upper parts of the stem enables the seeds to be dispersed separately—a vital matter for a wild plant. By contrast, modern cereal grasses have been selected, we believe by our ancestors, for tougher stems that do not break up when the grains (seeds) are ripe. The grains remain on the plant and can be harvested all together. This is true of wheat and barley but not of oats, in which it is hard to separate the grain from the chaff. Maize, on the other hand, has been so highly modified, probably under cultivation, that we have not been able to determine what its ancestor was. In fact it has been so modified that in its present form maize can hardly exist in the wild because it has lost its original power of seed dispersal.

Cultivated corn (maize) is easy to recognize, not only by its size, but by its seed-bearing organ, the ear. This is a highly specialized flower cluster enclosed in specialized leaves called husks. This cluster, when mature, bears several hundred naked seeds upon a rigid cob. No other plant has this type of seed-bearing organ. The unusual structure of a corn ear presents an interesting evolutionary problem. Domesticated corn has a low survival rate in nature, for, as we have said, it lacks a mechanism of seed dispersal. An ear of corn generally does not fall to the ground—it must be picked. However, if it does drop to or is left on the ground, scores of seedlings emerge. Usually, all of them die without ever reaching the reproductive stage since there is not enough moisture and soil nutrients for all of them to survive. It is believed that the ear of corn, a superb structure for bearing a large number of seeds, has come to have its present shape and structure as a result of humans deliberately selecting larger and fuller ears each year for their own use.

Out of the numerous species of higher plants, our ancestors selected 150 that they considered useful sources of food or fibers and they propagated and improved them. They did their job so well that we have not added any new crop to our agriculture since prehistoric times. At first they continuously looked for, found, and multiplied forms that possessed the desirable characteristics we associate with cultivated plants. The plants they chose were large and hardy; they lacked defensive structures such as hairs, spines, and thorns; they had shorter life spans. If their seeds were used for food, the plants with large seeds were selected. If their roots were used for food, the plants with fleshy roots were selected. If their fruits were used for food, the plants with tasty and juicy fruits were selected. This type of selection was the beginning of plant breeding. However, it was haphazard and slow since the desired characteristics were appearing by chance. Then, not wanting to wait for nature to give them the desired plants, our ancestors of long ago attempted to improve their crops by deliberately introducing into them the characteristics they wanted. They did this by a process of crossing and selection. They crossed a plant that had a desired characteristic with a plant of the same kind that did not, but which otherwise was a highly desirable plant.

A modern plant breeder paid tribute to his forerunners of years ago with these words:

> Without the primitive plant breeders who fashioned our food crops from the weeds about them, we would have none of the crops that supply most of our food to day. What kind of people were they? Probably most of them were women who gathered seeds and fruits while the men hunted. They knew nothing about genetics, sex in plants, metabolic pathways, DNA, RNA and the like. They were without microscopes, computers, or any of our sharply honed tools. They had no written language, no libraries, and probably no word for the introgression and recurrent selection that they practiced. What did they have? They had their plants. . . . They lived with them and knew them as few of us know ours. They had their hands and used them. They were motivated; their life depended on their own success. (Briggs and Knowles 1967, 3)

Our ancestors were both highly motivated and certainly as smart as we are, yet for centuries plant breeding remained more an art than a science. One reason is that until the seventeenth century little was known about sexuality in plants. Without such knowledge, modern breeding methods could not exist. Historians of biology credit Rudolph Jacob Camerarius, professor of natural history at the University of Tübingen in southern Germany, with being the first to demonstrate that plants as well as animals reproduce sexually. In a letter to his friend Professor Michael Bernard Valentin of the University of Giessen, he described experiments he had performed with spinach, hemp, hops, and maize that clearly demonstrated the functions of pollen and ovules. He found that when he cut off male flowers of these plants, no seeds were formed.

However, as is often the case in science, it took time for this discovery to be recognized. So it was with Camerarius's discovery that sexuality exists in plants. Joseph Gottlieb Kölreuter (1733–1806), one of the first "scientific" hybridizers (frequently mentioned by Mendel) and the first to point out the roles of insects and wind in transferring pollen, wrote: "From the 25th August 1694, when Camerarius wrote his letter concerning his experiments upon sex in plants, until September 1, 1761 there has been no real progress in the scientific knowledge which underlies plant breeding" (Hayes and Garber 1921, 5).

Progress was slow for two main reasons. First, in the eighteenth and even the nineteenth century many botanists refused to believe that sexuality exists in plants and so rejected the work of Camerarius, Kölreuter, and others on the sexuality of plants. The second reason was that most biologists at that time believed in a theory which today seems ridiculous and also seemed ridiculous to some scientists at that time. This was the theory of **preformation**, which assumed that the adult organism was already present in miniature in either the sperm or the egg. The growth of the embryo, triggered by fertilization, was simply the development of organs already present in microscopic form. That an embryo of a plant could be seen in the seed of the plant seemed to provide direct evidence of the correctness of this idea of preformation. However, such a theory denied the hereditary contribution of one of the parents, either the male or the female. Therefore, it could not explain the observations made by breeders for many years that offspring of a cross had some characteristics intermediate between their parents, some similar to those of one parent, and some similar to those of the other parent. These breeding results were always obtained in **reciprocal crosses**; that is, it made no difference which one was the pollen plant and which the seed plant. For this reason, nineteenth-century plant breeders became the proponents of the idea that sexuality exists in plants and the opponents of the preformation theory.

Further observation on the part of scientists using improved microscopes killed the theory of preformation, giving a boost to the contrasting theory of **epigenesis** which states that the fertilized egg gives rise to an embryo, then to the adult, by progressive differentiations of cells controlled by internal and external agents. Those agents are still not entirely known to us at this time. Moreover, sexuality in plants was finally confirmed in the second part of the nineteenth century, and plant hybridization had for the first time a scientific footing. Because of this the rate of crop improvement increased greatly throughout the nineteenth and twentiest centuries. Better varieties were developed and are still grown in many parts of the world because they have remained the best.

It is only recently that plant breeders have been able to create new varieties by intensively exploiting the idea that new desired characteristics are the result of gene mutations. Instead of waiting for those characteristics to appear in the wild, they developed methods to increase artificially the mutation rates of the genes responsible for those characteristics. They did this by applying certain chemicals and/or radiation from radium to X-rays. Even more recently, since 1980, plant breeders have been able to overcome the limits of sexual reproduction by the direct introduction of the desired genes into the cells of the plant itself. This type of plant breeding, genetic engineering, has opened up a whole new field of plant improvement. However, to be successful, modern plant breeders need more than a knowledge of botany, biochemistry, statistics, physics, and genetics. Like the ancient plant breeders, they need to know intimately the organism with which they work. As Barbara McClintock, the famous corn geneticist and winner of the 1983 Nobel Prize for medicine, said, "They [the plant researchers] have to have a feeling for the organism" (Keller-Fox 1983, 198). In other words, they have to be artists as well as scientists.

Botany of Peas

The people in the town of Brünn who knew and liked Father Mendel must have wondered what so occupied his attention in the monastery garden. Some knew that he spent much time there every day. He was growing peas with the vines held up off the ground by sticks and lengths of twine. He did not seem to care about the beauty of the flowers on the vines, for some said he was always pulling the flowers open and doing something to them. It was true that he was opening some of the buds on some of the vines, doing something to them, and then covering what was left of the buds with small cloth bags. Whatever it was he did, the plants formed pods and seeds inside the bags anyway.

If someone had asked Mendel what he was doing, he would probably have told them that opening the buds with his tweezers and taking out some of the parts that were there was all part of the process of artificially pollinating the buds to make them produce hybrid seed. To do this he had to transfer pollen from the flowers of one variety to one part of the flower of the other variety. Then he covered what was left of the bud with the cloth bag to make sure that no pollen from another plant could get onto the parts of that bud and spoil his experiment.

But there were other vines on which Mendel left the buds and flowers alone. These also produced pods and seeds, just as the other vines had. If he had been asked why he did not do anything to the buds of these vines, he probably would have explained that those vines were grown from pea seeds that he had produced the year before by artificial pollination. What he wanted to find out was what kinds of plants or seeds they would produce if they were allowed to grow naturally and pollinate themselves the way peas ordinarily did.

Explanations like these may have been good enough for the townsfolk of Brünn, but they are not good enough if we are to read Mendel's report of his experiments and understand what he wrote there. In order to really understand what he was doing, a more detailed knowledge of the botany of peas, including the structure of flowers, is needed. In what follows we have attempted to supply this knowledge without being too technical.

The garden pea, whose scientific name is *Pisum sativum*, is grown in many parts of the world and is consumed either as a fresh succulent vegetable or as a dried seed, usually used in soups. It is preeminently a cool-climate plant and will die if subjected for prolonged periods to temperatures above 77°F (25°C). On the other hand the plant can be frozen and, once thawed, will recover to continue its growth. Germination of the seed is rapid at temperatures above 60°F (15°C). Thus, in warmer regions peas are grown during the cooler part of the year.

We are not certain where garden peas first appeared, but it is highly probable that they originated in the mountain regions of southwestern Asia (Afghanistan and India). Alternatively, some botanists have suggested the garden pea was a natural hybrid between two wild pea species that grow in the eastern Mediterranean region. We know definitely that it has been cultivated for a very long time in India, starting possibly as early as 7000 to 6000 B.C. From India it was imported into Europe, where it has been hybridized so often both naturally and artificially that it has been transformed into many very different varieties. For example, the vines may range in height from 1½ to 7 feet. The flowers may be either white or purple and may be borne either at the end of a stem or along its sides. The seeds may be round, dimpled, or wrinkled. They may be green, yellow, or cream in color. To understand how such varieties can originate through hybridization it is necessary to have some knowledge of the structure of flowers and the mechanics of pollination.

In the first sentence of his paper *Experiments on Plant Hybrids,* Mendel referred to artificial fertilization of ornamental plants, but did not explain how this was done. When Mendel used the term *artificial fertilization* he actually meant *artificial pollination*. When he crossed peas of two different varieties to form a hybrid, he did so by artificial pollination. No matter whether pollination is accomplished naturally or artificially, **fertilization** has always followed a natural, predetermined course. While Mendel could deliberately control the pollination of his pea flowers, he was not able to control the process of fertilization. This

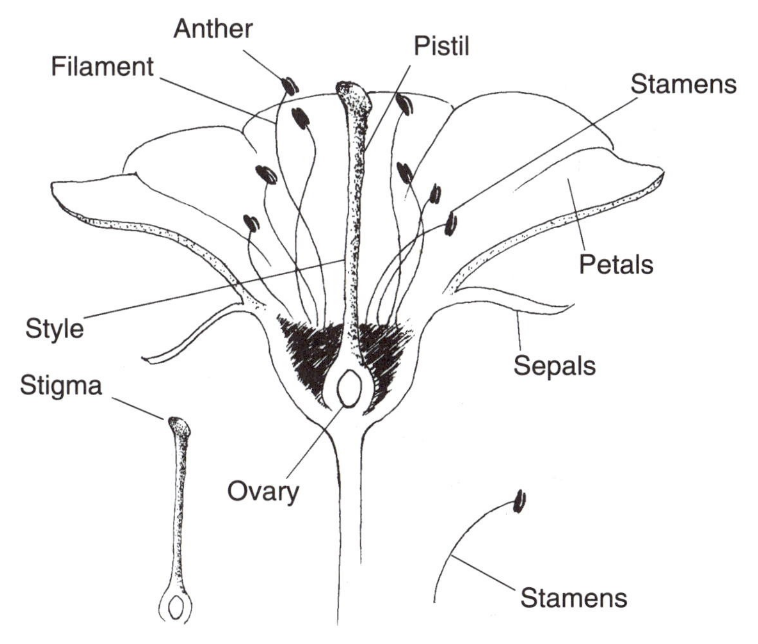

Fig. 3.1. Diagram of a Flower

depended, as it still does, on biological and chemical processes beyond human control.

Because the process of fertilization in flowering plants is complex and does not bear directly on an understanding of Mendel's paper, we have placed the discussion of this process in appendix A. Here we have restricted the discussion to the information readers need to understand what Mendel wrote.

Structure of a Flower

The first parts of the flower, beginning outside and at its base, are the **sepals** (fig. 3.1). These, taken together, form the **calyx**, which usually encloses the other parts of the flower when it is still closed. The sepals are usually green, but in some plants they are colored. The next floral parts as we go toward the center of the flower are the **petals**. Taken together, these make up the structure known as the **corolla**. The petals are generally the more or less brightly colored parts attractive both to us and to insects. They are what most people think of as flowers.

Next, inside the petals and often extending above them are the **stamens**. These are the male parts of the flower. Each stamen consists of a long, more or less slender **filament** or stalk and, at its outer end, a somewhat larger structure, the **anther**. The anther contains the pollen grains, which carry the male contribution in fertilization.

Located at the center of the flower is the **pistil**, which is the female part of the flower (some flowers have more than one pistil). The pistil usually has a somewhat enlarged section near the base called the **ovary**. Inside this are the **ovules**, which contain the egg cells along with other materials. If the egg cells are fertilized the ovules develop into seeds.

Extending upward from the ovary is a slender column, the **style**. At the top of this, there is generally an enlarged structure called the **stigma**. This structure is adapted to receive and hold pollen grains.

The stamens and pistils are regarded as the essential parts of the flower. However, in certain species, the flowers may lack either one or the other of these structures. These **imperfect flowers** may occur on different individual plants or both types may occur in one plant.

Pollination

The first step in fertilization is *pollination*, the transfer of pollen grains from an anther to a stigma. This transfer may take place in any one of a number of different ways, either naturally or artificially. In natural pollination the transfer agent may be gravity, wind, or the activities of insects. In artificial pollination, pollen is often transferred with a small brush, such as is used for painting with watercolors. The pollen grains are picked up on the brush and then dusted or "painted" onto a stigma. Alternatively, the pollen grains may be shaken from one or more anthers into a bag, which is then inverted over the pistil or pistils of the same or another flower and shaken. The pollen grains then fall onto the stigma(s). This is roughly the procedure used in artificial pollination of corn. If the pollen from an anther of a flower is transferred to the stigma of the same flower, or to another flower on the same plant, the flower receiving the pollen is said to be **self-fertilized** or **selfed**. If, on the other hand, the pollen from the anther of one variety is transferred to the stigma of a flower of a different variety, the flower is said to be cross-fertilized or crossed.

Pea flowers are characteristic of a large number of plants belonging

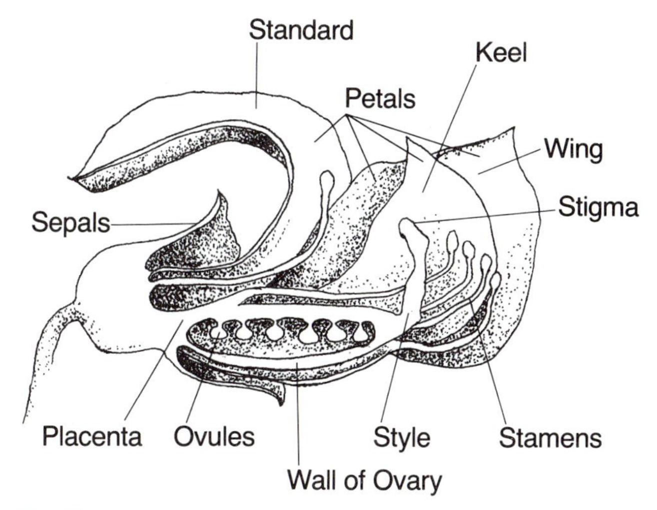

Fig. 3.2. Pea Flower

to the family *Papillonaceae*, so called because the flowers tend to look like butterflies. They are said to be **complete flowers** because they have all the usual structural parts: sepals, petals, stamens, and pistils. They are also said to be **perfect flowers** because each one has all the usual reproductive parts (fig. 3.2).

In the pea flower, the sepals, which are five in number, are united along their edges only at the stem end. Only two of the five petals are united, forming the structure known as the **keel**. The other three petals are separated. The larger one is called the **standard**, the other two are called the **wings**. In the bud stage the standard is folded down and wraps around the other petals.

The lower portions of the filaments of nine of the ten stamens are fused to form a tube over nearly the entire length of the ovary. The tenth stamen is free throughout its length. In the very young flower, the filaments are shorter than the style; but by the time the anthers release pollen, the filaments have forced the anthers tightly into the tip of the keel and against the style.

The pistil is attached by a stalk to the middle of the bottom of the sepals. The green ovary, which may bear up to thirteen ovules but generally bears only eight, is located in the swollen base of the pistil. The slightly flattened, cylindrical style extends from the top of the ovary and is bent at nearly a right angle to it. The stigma, on whose sticky surface the pollen grains fall in pollination, is located at the tip of the style.

Peas offer two advantages in hybridizing experiments. One is that their flowers are large and easy to work on. The other is that, due to the special structure of their flowers, pollination takes place before the flowers open. Peas are therefore naturally **self-pollinated** and there is a very low frequency of accidental **cross-pollination**.

A pea plant, like any flowering plant, starts as a fertilized egg, or **zygote**. Fertilization brings a sudden end to the functions of the flower. The sepals, petals, and stamens wither, but the ovary develops into a fruit. Inside the ovary the ovules, which contain the developing embryos, mature into seeds.

The fruit of the pea plant is called a **pod**. It is really a mature ovary with remnants of style and stigma sometimes persisting at the tip of the ovary. The pod is made up of two walls. Garden peas are divided into two groups on the basis of whether the walls of their pods do or do not contain a layer of tough fiber called *parchment*: the pods of the shelling forms contain parchment and are not edible; the pods of sweet forms do not contain parchment and are edible.

The shape of a pea pod is variable. Pea pods can be straight, saber-shaped or sickle-shaped, smooth or constricted (fig. 3.3). The color of the immature pod is also variable. It may be yellow, light green, or dark green. The color of the mature pod shows less variability. It may be light yellow, brown, or rarely, violet brown. The number of seeds in a pod is also variable. It may be as low as three and as high as twelve.

Since Mendel's experiments involved traits such as seed shape, cotyledon colors, and seed coat colors, it will be helpful for the reader to know something about the structure of a pea seed and how the seed develops after fertilization. The pea seed consists of two main parts: the **embryo**, which is the future plant, and the **seed coat**. The main parts of the embryo are (figs. 3.4 and 3.5):

1. The **plumule**, which is a rudimentary shoot
2. The **cotyledons**, which are the first leaves
3. The **hypocotyl**, to which the cotyledons and the upper part of the rudimentary shoot are attached
4. **The radicle**, which is the rudimentary root

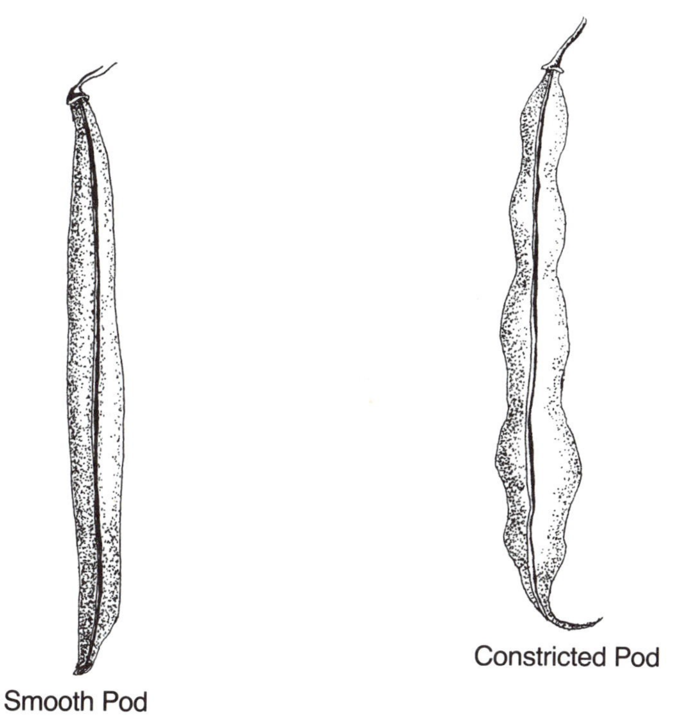

Fig. 3.3. Pea Pod Shapes

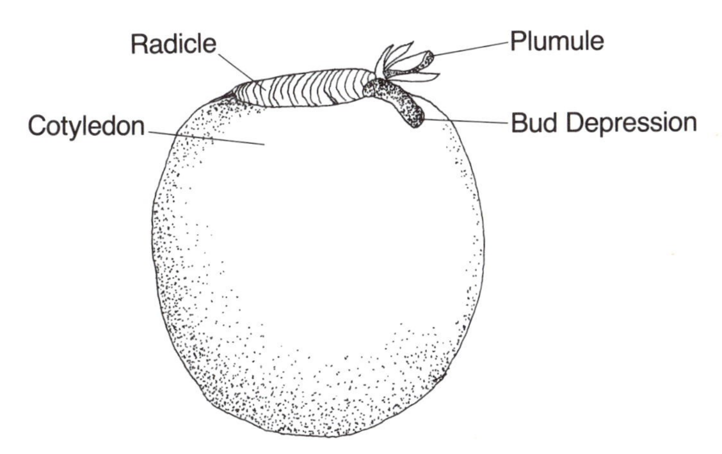

Fig. 3.4. Germinating Seed (only one cotyledon shown)

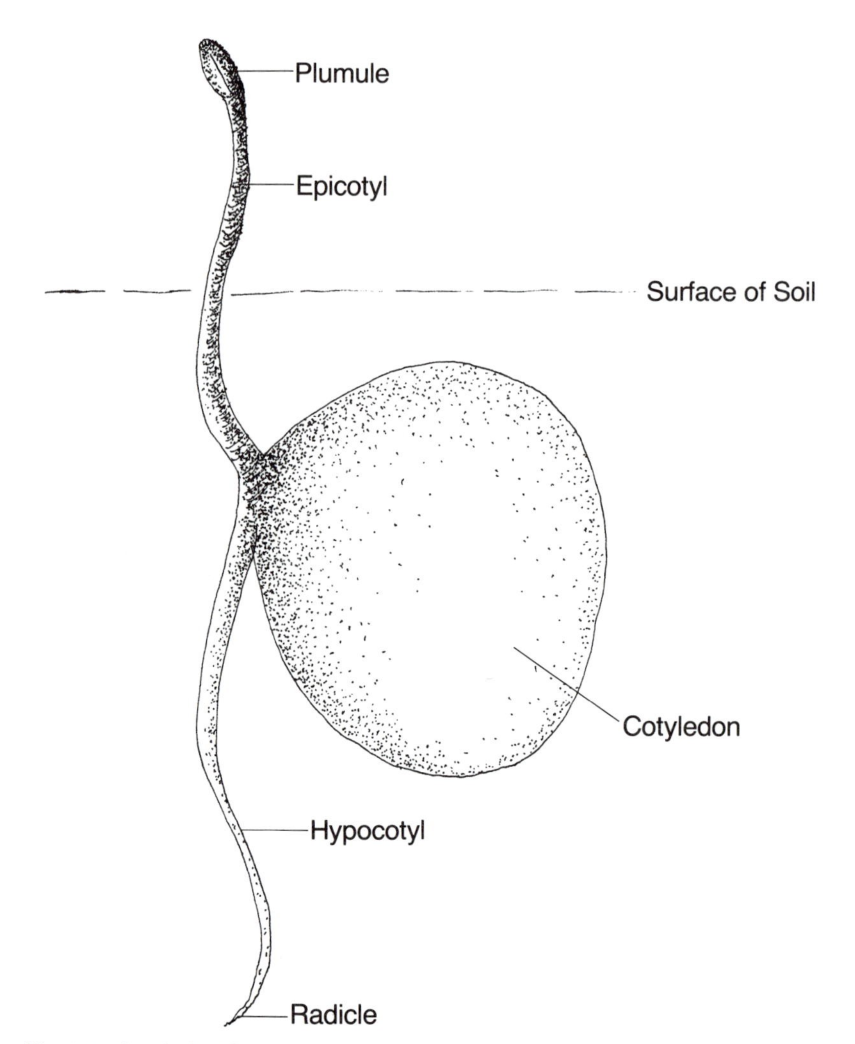

Fig. 3.5. Seed after Germination

The plumule, hypocotyl, and radicle together form the axis of the embryo. At germination the plumule develops into the shoot that rises above the ground and matures into the visible plant, and the radicle develops into the primary root below the ground. The cotyledons are specialized leaves attached to the axis of the embryo. They are unlike ordinary leaves because they are food-storing organs for the developing seedling. The growing pea seedling gets its food from the cotyledons, which remain below the ground and finally disintegrate.

The seed coat is not part of the embryo, but is part of the plant that bore the seed. It may be transparent (white) or it may be brown. If the seed coat is transparent, it will not influence the color of the seeds. In varieties with transparent seed coats (white), seed color depends to a great extent on the color of the cotyledons, either green or yellow, which can be seen through the translucent and almost colorless seed coat. If the seed coat is colored, it is generally reddish, brown, or purple black. In this case, it will be opaque and the color of the cotyledons cannot be seen. We shall see later that an understanding of the anatomical origin of the pea seed coat is very important in understanding Mendel's experiments.

A seed can remain dormant for a long period, sometimes for years. But sooner or later, if the conditions are favorable, seed germination will take place, growth will be resumed, and a new plant generation will be on its way. When the seed germinates, the hypocotyl elongates, pushing the plumule to the surface and giving rise to the stem and leaves of the young plant. The radicle develops into a primary root and, later, the entire root system. The cotyledons either shrivel up when the food they contain has been absorbed by the developing embryo, which is now ready to manufacture its own food by photosynthesis, or they are transformed into true leaves.

You now have the minimum basic information about plant breeding and the botany of peas needed to begin the reading of Mendel's paper, which follows next in this text.

Experiments on Plant Hybrids by Gregor Mendel

Text and Interpretation

General Introduction

Gregor Mendel's short treatise "Experiments on Plant Hybrids" is one of the triumphs of the human mind. It does not simply announce the discovery of important facts by new methods of observation and experiment. Rather it is an act of highest creativity, it presents these facts in a conceptual scheme which gives them general meaning. Mendel's paper is not solely a historical document. It remains alive as a supreme example of scientific experimentation and profound penetration of data. It can give pleasure and provide insight to each new reader—and strengthen the exhilaration of being in the company of a great mind at every subsequent study.

Stern and Sherwood 1966, 5

Elaborate apparatus plays an important part in the science of today, but I sometimes wonder if we are not inclined to forget that the most important instrument in research must always be the mind of man.

W. B. Beveridge, The Art of Scientific Investigation

Mendel's paper, *Experiments on Plant Hybrids*, is a long one. Mendel himself divided it into eleven sections. The first seven sections are easier to understand than the last four. As we follow Mendel's paper from the statement of the objective of his experiments to the explanation of their results and their consequences, we can distinguish various levels of increasing complexity in the way he was thinking about his material. These levels, which are reflected in the way he wrote about his experiments and their results, are the following:

1. Qualitative description in natural language
2. Quantitative description
3. Constant relationships, partly expressed in words and partly in numbers
4. Empirical laws, expressed in words and symbols
5. Theory formation

The meanings of these levels and their relation to the sections will become clear as you work your way through Mendel's text. To help make this text more accessible to you, we have, for each of the first seven sections, written a short introduction telling you what the section is about. This is followed by Mendel's text. Then an explanation of that section with appropriate comments follows. We have subdivided the last four sections, which are harder to understand, into two or more parts using the same format for each section.

§1

Introductory Remarks

At the beginning of this first section Mendel tells us very succinctly why he got involved in carrying out the experiments he is reporting. He was led to do them by results of previous hybridizing experiments with ornamental plants, the objective of which was to obtain varieties having new colors. He does not tell us whether they were successful or not, but he tells us that the same hybrid forms reappeared with "striking regularity." These two words are very important because Mendel believed, as every scientist does, that there is regularity in nature. For scientists, it is a fundamental assumption that the universe is orderly and can be understood. They are sure that, given time, they will be able to perceive this order in all things, that they will be able to see constant relationships. These constant relationships are designated as **laws** and are often identified by the name of the scientist who made the discovery.

Let us go back to Mendel's expression, "striking regularity." What did he mean by it? Let us suppose, for example, that the experiments had been done with carnations, crossing red ones with white ones, and that each time the experiment was done, only pink carnations were produced. Since pink carnations were already known, the experiments would not produce anything new. In this they would be a failure. But they did produce those pink carnations every time. Probably, this was not very exciting at first. But somewhere in Mendel's memory that regularity of behavior of flower color rang a little bell. Regularity of behavior is regarded in the physical sciences as evidence that the observed behavior is governed by a natural law. Would this be true in biology as well? Perhaps yes, answered Mendel, the Augustinian monk and priest. Since God created the whole world, why should natural laws exist only in physics and chemistry? More likely, they exist in biology also, but no one had looked for them in the right way. If that were the case, then that regularity of behavior in the hybrids produced in the attempt to form new color variants might be a clue. In this context, what had seemed to be failed experiments now became a possible clue to a whole new area of investigation: the search for the law governing the formation and development of hybrids over several generations.

The same type of clue Mendel used was also used by the fictitious detective Sherlock Holmes in solving the mystery of the race horse Silver Blaze. The horse had been let out of the barn at night by an intruder, and its trainer had been found dead on the moor near the barn. Two boys with their dog were sleeping in the barn that night and heard nothing. Therefore, it was possible for Holmes to infer that the dog had not barked because it knew the intruder. So, by like reasoning as in the ornamental hybrid experiment, something that did not happen—no new color variants could be obtained—led to the solution of why something happened.

Perhaps it seems strange that Mendel knew something about natural laws in physics and chemistry. After all, Mendel has been known for a long time mainly as the discoverer of the laws of heredity and that involves biology, not physics and chemistry. But, if you remember from the brief biography of Mendel earlier in this text, he spent much of his life as a teacher, not of biology, but of experimental physics. And you remember that he spent a great deal of his two years at the University of Vienna studying physics, chemistry, and mathematics. At that time the sciences of physics and chemistry were quite advanced as compared to biology. They laid great stress on the importance of discovering quantitative natural laws and, once these had been discovered, using them to understand nature. Training to be a physical scientist was training in the process of discovering and using natural laws. In his own training Mendel would have been aware of how few such natural laws there were in the biological as compared to the physical sciences. And he would have been thoroughly indoctrinated in the idea that the highest contribution a scientist can make to science is the discovery of one of these natural laws. Then, if Mendel was successful, he would join Robert Boyle, J. Charles, Joseph Gay-Lussac, Isaac Newton, Georg S. Ohm and others as the discoverer of a law that would bear his name. We believe that it was this Vienna experience that sensitized Mendel to the importance of that invariant relationship (the regularity of appearance of hybrid forms) and led him to refer to it as "striking." And so he was led to carry out "further experiments whose task it was to follow the development of hybrids in their progeny."

When a scientist has decided to carry out experiments on a particular problem, he or she first conducts a survey of the literature (books and technical papers that contain reports of what earlier researchers have found when doing similar experiments). Once the survey has been completed, the investigator has to decide how this information affects the general plan, then formulates a broad outline of what kind of data will have to be obtained to carry out this plan. This is what Mendel does in the rest of this section.

Mendel's Text

1 Artificial fertilization undertaken on ornamental plants to obtain new
color variants initiated the experiments to be discussed here. The striking
regularity with which the same hybrid forms always reappeared whenever
fertilization between like species took place suggested further experiments
5 whose task it was to follow the development of hybrids in their progeny.
Numerous careful observers, such as Kölreuter, Gärtner, Herbert,
Lecoq, Wichura, and others, have devoted a part of their lives to this
problem with tireless persistence. Gärtner, especially, in his work "Die
Bastarderzeugung im Pflanzenreich" (Hybrid Production in the Plant
10 Kingdom) has recorded very estimable observations, and Wichura has
very recently published the results of his thorough investigations of
willow hybrids. That no generally applicable law of the formation and
development of hybrids has yet been successfully formulated can hardly
astonish anyone who is acquainted with the extent of the task and who can
15 appreciate the difficulties with which experiments of this kind have to
contend. A final decision can be reached only when the results of
detailed experiments from the most diverse plant families are available.
Whoever surveys the work in this field will come to the conviction that
among the numerous experiments not one has been carried out to an extent
20 or in a manner that would make it possible to determine the number of
different forms in which hybrid progeny appear, permit classification of
these forms in each generation with certainty, and ascertain their
numerical interrelationships. It requires a good deal of courage indeed
to undertake such a far-reaching task; however, this seems to be the one
25 correct way of finally reaching the solution to a question whose
significance for the evolutionary history of organic forms must not be
underestimated.

This paper discusses the attempt at such a detailed experiment. It was expedient to limit the experiment to a fairly small group of plants, and
30 after a period of eight years it is now essentially concluded. Whether the plan by which the individual experiments were set up and carried out was adequate to the assigned task should be decided by a benevolent judgment.

Interpretive Comments

The opening paragraph of these introductory remarks (lines 1–5) appears to be quite simple and straightforward as we first read it. In part this is so, but careful examination turns up a whole host of interesting questions. Take the first sentence for example. Mendel writes that certain experiments started him on the research that he is reporting in this paper. But who did these experiments? Did he do them or did someone else? What were the ornamental plants? He does not tell us, so we do not know. Were the experiments successful? Were new color variants produced or not? Again we do not know. It seems reasonable that if the experiments had been successful he would have said so, but he did not. As you can see, the first sentence of this paragraph, simple and direct as it is, offers all kinds of opportunities for scientific detective work.

Mendel covers the summary of his literature review in lines 6–12. When he reviewed the work of other hybridizers he found, as he wrote in lines 18 through 23, that none of them had carried out their experiments in such a way as to provide the kind of data he needed.[1] For, in order to find a law or laws responsible for the behavior of hybrids, Mendel foresaw that he would need quantitative data of a certain kind. However difficult the task of designing the experiment and obtaining and analyzing the data might be, Mendel saw, as he wrote, no other way to go (lines 23–27).

In the last short paragraph of this section (lines 28–32) Mendel notes that it has taken him eight years to design the necessary experiments, carry them out, and interpret the evidence he has collected and that he would leave it to the reader to judge how well he had succeeded in what he had set out to do.

Notes

Mendel's paper was first read at two consecutive meetings of the Natural Science Society of Brünn, February 8 and March 8, 1865. It was published in

1866 in *Verhandlungen des Naturforschenden Vereines in Brünn* 4, Abhandlungen 3–47.

The first matter of significance here is the title of the paper, which is important both for what it does say and for what it does not. It indicates that Mendel is reporting the results of his experiments on plant hybrids, but does not say that the experiments were about the inheritance of certain characters in plants, nor does it say anything about being concerned with finding the laws of inheritance. It is important to keep these points in mind because, as you know from the introduction, the traditional account has always taken the view that the paper is basically about inheritance and its laws.

1. Mendel cites the following hybridizers: J. G. Kölreuter, a German botanist and hybridizer who published numerous papers on hybrid plants between 1761 and 1796; C. G. Gärtner, a German botanist and hybridizer who did extensive research on the structure of pollen grains and upon the process of pollination in a wide group of plants and whose most important paper on hybrids was published in 1838; W. Herbert, an English botanist and horticulturist who wrote papers between 1819 and 1847, mostly on hybrids of vegetables; H. Lecoq, a French botanist who wrote two main books on the reproduction and hybridization of plants, the first one in 1827 and the second one in 1846, with a revised edition in 1862; and M. E. Wichura, a German botanist and hybridizer whose book on plant hybrids was published in 1865. Wichura also wrote a preliminary report on willow hybrids in 1854. Obviously, this is the one to which Mendel referred.

§2

Selection of Experimental Plants

Remember that Mendel's problem was to find the law that governs the formation of hybrids and their offspring over several generations. To do so he would have to make hybrids and produce their offspring in such a way or ways that no foreign kinds of plants could slip into the results and spoil them. In addition, he would have to keep very complete and exact records of what he did and what the results were. If this was what he was going to try to do, it would be very important to pick out of all the many species and varieties that were available to him just the right kind of plants to use. That is what this whole section is about, the selection of his experimental plants. We will discuss this process in the comments on the text itself, but first we want to look briefly at some other examples where selection of the right experimental organism made critical differences in several important biological experiments leading to breakthroughs in science.

In addition to Mendel's choice of the garden pea in his studies of hybrids in the 1860s, we can think of two examples: the first is the selection of the breadmold, *Neurospora crassa*, by G. W. Beadle and E. L. Tatum, in their successful attempt in the 1940s to study what genes do and how they do it; and the second is the choice of **bacteriophages** (the viruses that attack bacteria) by geneticists in the 1950s and 1960s in their study of the nature of the gene. In each instance, because of their specific properties the chosen organisms gave the researchers just the tool needed to solve their specific problem. These organisms have several things in common that facilitate research: they can be grown easily, they reproduce rapidly, and they reproduce in large numbers. Think of the usefulness of working with bacteria or viruses, which can reproduce in less than an hour and the disadvantage of working with trees, whose reproductive cycle is not less than twenty years. Think of the usefulness of working with bacteria, or even peas, which produce a large number of offspring in one season, and the drawback of working with elephants, which produce only one calf every two years.

Mendel's Text

1 The value and validity of any experiment are determined by the suitability of the means used as well as by the way they are applied. In the present case as well, it can not be unimportant which plant species were chosen for the experiments and how these were carried out.

5 Selection of the plant group for experiments of this kind must be made with the greatest possible care if one does not want to jeopardize all possibility of success from the very outset.

The experimental plants must necessarily

1. Possess constant differing traits.

10 2. Their hybrids must be protected from the influence of all foreign pollen during the flowering period or easily lend themselves to such protection.

3. There should be no marked disturbances in the fertility of the hybrids and their offspring in successive generations.

15 Contamination with foreign pollen that might take place during the experiment without being recognized would lead to quite erroneous conclusions. Occasional forms with reduced fertility or complete sterility, which occur among the offspring of many hybrids, would render the experiments very difficult or defeat them entirely. To discover the 20 relationships of hybrid forms to each other and to their parental types it seems necessary to observe *without exception all* members of the series of offspring in each generation.

From the start, special attention was given to the *Leguminosae* because of their particular floral structure. Experiments with several members of 25 this family led to the conclusion that the genus *Pisum* had the qualifications demanded to a sufficient degree. Some quite distinct forms of this genus possess constant traits that are easily and reliably distinguishable, and yield perfectly fertile hybrid offspring from reciprocal crosses. Furthermore, interference by foreign pollen cannot 30 easily occur, since the fertilizing organs are closely surrounded by the keel, and the anthers burst within the bud; thus the stigma is covered with

pollen even before the flower opens. This fact is of particular importance. The ease with which this plant can be cultivated in open ground and in pots, as well as its relatively short growth period, are
35 further advantages worth mentioning. Artificial fertilization is somewhat cumbersome, but it nearly always succeeds. For this purpose the not yet fully developed bud is opened, the keel is removed, and each stamen is carefully extracted with forceps, after which the stigma can be dusted at once with foreign pollen.

40 From several seed dealers a total of 34 more or less distinct varieties of peas were procured and subjected to two years of testing. In one variety a few markedly deviating forms were noticed among a fairly large number of like plants. These, however, did not vary in the following year and were exactly like another variety obtained from the same seed dealer;
45 no doubt the seeds had been accidentally mixed. All other varieties yielded quite similar and constant offspring; at least during the two test years no essential change could be noticed. Twenty-two of these varieties were selected for fertilization and planted annually throughout the entire experimental period. They remained stable without exception.

50 Their systematic classification is difficult and uncertain. If one wanted to use the strictest definition of species, by which only those individuals that display identical traits under identical conditions belong to a species, then no two could be counted as one and the same species. In the opinion of experts, however, the majority belong to the species *Pisum*
55 *sativum*; while the remaining ones were regarded and described either as sub-species of *P. sativum*, or as separate species, such as *P. quadratum*, *P. saccharatum*, and *P. umbellatum*. In any event, the rank assigned to them in a classification system is completely immaterial to the experiments in question. Just as it is impossible to draw a sharp line between species
60 and varieties, it has been equally impossible so far to establish a fundamental difference between the hybrids of species and those of varieties.

Interpretive Comments

Here again Mendel opens his comments with one of those very compressed, somewhat cryptic remarks that require some detective work. Probably the best way to get at what he means in his opening remarks is to analyze what he wrote and then reassemble it like a jigsaw puzzle.

It seems to us that what he meant by the "value" of an experiment was whether it seemed likely to produce results of real significance. He wrote that the information gained must be valid, that is, trustworthy, reliable. If it does not possess this characteristic it has no real importance; it is not useful. By "the means used" in the experiment he meant the methods the experimenter employs in producing the data obtained and how they are applied—skillfully, or are important sources of error and difficulty just ignored? Finally, all of these matters are strongly influenced by the kind of problem the experimenter is investigating.

In lines 1 and 2 Mendel indicates that he is very much aware of the importance of choosing the right experimental methods and using them in the right way. As we mentioned above, this sensitivity was undoubtedly a consequence of his experiences as a special student at the University of Vienna during 1851 and 1852. One of the textbooks he used was written by two of his professors. The following quotation will show you where he might have gotten the idea we have noted in lines 1 and 2: "According to Baumgartner and Ettinghausen, the preparation of suitably organized experiments is the surest means of learning the ways of nature in its many aspects, of finding the keys to the laws, and of understanding the interrelation of phenomena" (Orel 1984a, 30).

However, Mendel was thinking about his experiments not only as a physicist but also as a botanist and a hybridizer. He recognized that in order to get the quantitative data needed in his search for the law or laws of hybrids, he had to have plants of the right type. In the statements numbered 1, 2, 3 in lines 9–14 he lays down the specific attributes that his plants must have. Statements 2 and 3 are self-explanatory, but statement 1 needs additional comments. In that statement he refers to "constant differing traits." In modern usage a plant **trait** is a distinguishing feature of a plant such as seed shape, seed color, stem length, or position of flowers. These traits may be present in only one form or in several. Mendel selected varieties of peas showing each trait in two clearly different forms. For example, the seeds could be either round or angular and they could be either green or yellow. In modern terms these different forms are referred to as the **characters** of the trait. We shall maintain this distinction as consistently as possible, but you should be aware that Mendel was not entirely consistent in his usage. Usually, however, a careful examination of what he is writing about will enable you to distinguish which of the modern terms is the appropriate one. From this brief commentary you can see what Mendel meant by "different traits." Some of the parental characters he used reappeared in the hybrids unchanged. Others, though they did not appear in the hybrids, reappeared unchanged in the offspring of the hybrids.

Mendel's choice of the garden pea to use in his studies of the behavior of hybrids was wise and fortunate for the following reasons: (1) true-breeding varieties possessing constant differentiating characters are easily observable and easily obtainable; (2) there is very little cross-pollination in peas; and (3) hybrids and their offspring suffer no marked disturbance in their high fertility in successive generations. We should note that quite a few hybridizers before Mendel had used peas for many of the same reasons that led him to use them. Let us develop these ideas further.

Peas belong to the large family of Leguminosae (mentioned in line 23), which in more familiar terms are called legumes. The group includes peas (species name *Pisum*), beans (species name *Phaseolus*), and lentils (species name *Lens*). Peas are **annual** plants—plants that grow from seed to seed in one year—which can be grown and crossed easily. Their flowers are so constructed that pollen from a flower normally falls on the stigma of that flower and thus effects self-fertilization. Cross-fertilization is rare without human intervention. Through many generations of natural self-fertilization, garden peas have developed into a number of true-breeding varieties with well-defined characteristics.

Natural self-pollination in plants such as peas makes the task of a hybridizer, such as Mendel, much easier. First, to make a cross between two different pea plants, the hybridizer can open a flower bud of one plant (which will serve as the seed parent of the cross) and remove the stamens before any pollen has been shed (thus preventing self-pollination). When the bud has been opened, the immature stigma, which is now exposed, must be protected with a cover (such as a small paper bag) for the few days during which it becomes ready to accept and retain pollen. Then the hybridizer removes the protective cover and places on the stigma some pollen from the plant that he or she wishes to use as the male parent of the desired cross. It is, of course, necessary to guard this artificially fertilized flower against contamination by pollen of unknown origin, which might be brought to it by wind, insects, or the hybridizer's hands or instruments. Though Mendel did not mention this, he must have used some kind of bag to protect his cross-pollinated flowers.

Second, if Mendel wished to determine the kind of progeny that would appear among the offspring of a hybrid, all he had to do was allow the flowers of the hybrid to fertilize themselves naturally. This permitted him to obtain a very large number of seeds without much work on his part. As you will see later, the large number of offspring that he obtained allowed him to be successful in making a correct succesful interpretation of his results.

Garden peas presented Mendel with some additional advantages. A single plant does not take up much space. Because of this, Mendel was able to plant a considerable number of seeds in the small plot he had in the monastery. Also, the plants mature in a short time—an advantage in Moravia, which has a short growing season.

Having made the decision to use peas as his test organism, Mendel obtained seeds of thirty-four different varieties from a number of different sources (line 40). In order to assure himself that each form of plant he wanted to use in his experiments met the criteria he had established, he grew them in test beds for two years with the results he notes in lines 41–49. From these he selected twenty-two varieties, which he used during the course of his experiments.

In this paragraph (lines 40–49) Mendel uses the word **variety** for his breeding lines, and yet in his "Introductory Remarks" he wrote about hybrids between species. There is a distinction between **species** and varieties, and Mendel alludes to this in the last paragraph of this section. So, in order to understand the importance of the last paragraph of this section (lines 50–62), we must pause to fill in some background information on species, varieties, and classification. When Mendel wrote that the systematic classification of his pea plants was uncertain, he was saying that it was not clear whether they were all different varieties of one species or were different species. In modern terminology, a species is composed of plants or animals that are able to breed successfully with one another. They share certain distinguishing traits in common, which serve as the bases for classifying them as members of that species. In other traits, members of a species may show considerable differences without affecting their classification. These differences may characterize subgroups, which are considered varieties of the species. However, so long as members of the several varieties remain

interfertile, they are all considered to be members of one species. If, on the other hand, plants or animals are not interfertile, they are considered to be members of different species.

In Mendel's day these distinctions between species and varieties were not as well worked out as they are now. However, all of Mendel's varieties of peas were interfertile and were therefore members of only one species. From his viewpoint, so long as the plants met his criteria, their formal classification was of little or no importance.

§3

Arrangement and Sequence of Experiments

Having made clear to us why he selected peas as his experimental plants, Mendel turned in this section to describing the various traits of peas he has selected to use in his experiments. As he points out, not all available traits are useful, but only those that show clearly defined and sharply contrasting characters. Having done this he describes his initial experiments, discussing some of the problems he met and how he solved them. One of these problems was possible pollination of his experimental plants in the garden with foreign pollen carried to them by insects. In order to detect such contamination he placed some plants in pots in a greenhouse while they were in flower, thus shielding them from insects. These plants provided results of known purity against which to judge results obtained in the garden. Mendel correctly refers to these protected plants as controls. Here they served as controls for a recognized single source of possible error and not for a group of possible but unknown sources of errors, as in the case when a control group of plants is used in some modern experiments. However, it is but one of many evidences of the knowledge, care, and skill Mendel brought to these experiments.

Mendel's Text

1 When two plants, constantly different in one or several traits, are
crossed, the traits they have in common are transmitted unchanged to the
hybrids and their progeny, as numerous experiments have proven; a pair of
differing traits, on the other hand, are united in the hybrid to form a
5 new trait, which usually is subject to changes in the hybrid's progeny. It
was the purpose of the experiment to observe these changes for each pair
of differing traits, and to deduce the law according to which they appear in

successive generations. Thus the study breaks up into just as many separate experiments as there are constantly differing traits in the
10 experimental plants.

The various forms of peas selected for crosses showed differences in length and color of stem; in size and shape of leaves; in position, color, and size of flowers; in length of flower stalks; in color, shape, and size of pods; in shape and size of seeds; and in coloration of seed coats and
15 albumen. However, some of the traits listed do not permit a definite and sharp separation, since the difference rests on a "more or less" which is often difficult to define. Such traits were not usable for individual experiments; these had to be limited to characteristics which stand out clearly and decisively in the plants. The result should ultimately show
20 whether in hybrid unions the traits all observe concordant behavior, and whether one can also make a decision about those traits which have minor significance in a classification.

The traits selected for experiments relate:

1. *To the difference in the shape of the ripe seeds*. These are either
25 round or nearly round, with depressions, if any occur on the surface, always very shallow; or they are irregularly angular and deeply wrinkled (*P. quadratum*).

2. *To the difference in coloration of seed albumen* (endosperm). The albumen of ripe seeds is either pale yellow, bright yellow and orange, or has a more or less intense green color. This color difference is easily
30 recognizable in the seeds because their coats are transparent.

3. *To the difference in coloration of the seed coat*. This is either white, in which case it is always associated with white flower color; or it is grey, grey-brown, leather-brown with or without violet spotting, in
35 which case the color of the standard is violet, that of the wings is purple, and the stem bears reddish markings at the leaf axils. The grey seed coats turn black-brown in boiling water.

4. *To the difference in shape of the ripe pod*. This is either smoothly arched and never constricted anywhere, or deeply constricted between the
40 seeds and more or less wrinkled (*P. saccharatum*).

5. *To the difference in color of the unripe pod*. It is either colored light to dark green or vivid yellow, which is also the coloration of the stalks, leaf-veins, and calyx.[1]

1. One variety has a beautiful brownish-red pod color which tends to a violet and blue around the time of ripening. The experiment with this trait was started only in the past year.

6. *To the difference in position of flowers*. They are either axillary, that is, distributed along the main stem, or they are terminal, bunched at the end of the stem and arranged in what is almost a short cyme; if the latter, the upper part of the stem is more or less enlarged in cross section (*P. umbellatum*).

7. *To the difference in stem length*. The length of the stem varies greatly in individual varieties; it is, however, a constant trait for each, since in healthy plants grown in the same soil it is subject to only insignificant variations. In experiments with this trait, the long stem of 6 to 7′ was always crossed with the short one of ¾′ to 1½′ to make clear-cut distinction possible.

Each of the two differing traits listed above as pairs were united by fertilization.

For the

1st	experiment	60	fertilizations	on	15	plants	were	undertaken.
2nd	"	58	"	"	10	"	"	"
3rd	"	35	"	"	10	"	"	"
4th	"	40	"	"	10	"	"	"
5th	"	23	"	"	5	"	"	"
6th	"	34	"	"	10	"	"	"
7th	"	37	"	"	10	"	"	"

From a fairly large number of plants of the same kind only the most vigorous were chosen for fertilization. Weak plants always give uncertain results, because many of the offspring either fail to flower entirely or form only few and inferior seeds even in the first generation of hybrids, and still more do so in the following one.

Furthermore, in all experiments reciprocal crosses were made in such a manner that that one of the two varieties serving as seed plant in one group of fertilizations was used as pollen plant in the other group.

The plants were grown in garden beds—except for a few in pots—and were maintained in their natural upright position by means of sticks, twigs, and taut strings. For each experiment a number of the potted plants were placed in a greenhouse during the flowering period; they were to serve as controls for the main experiment in the garden against possible disturbance by insects. Among the insects that visit pea plants the beetle *Bruchus pisi* might become dangerous to the experiment should it appear in fairly large numbers. It is well known that the female of this species lays her eggs in the flower and thereby opens the keel; on the tarsi of one specimen caught in a flower some pollen cells could clearly

be seen under the hand lens. Mention must also be made here of another circumstance that might possibly lead to the admixture of foreign pollen.
85 For in some rare cases it happens that certain parts of an otherwise quite normally developed flower are stunted, leading to partial exposure of the fertilization organs. Thus, defective development of the keel, which left pistil and anthers partly uncovered, was observed. It also sometimes happens that the pollen does not fully mature. In that event the pistil
90 gradually lengthens during the flowering period until the stigma protrudes from the tip of the keel. This curious phenomenon has also been observed in hybrids of *Phaseolus* and *Lathyrus*.

The risk of adulteration by foreign pollen is, however, a very slight one in *Pisum*, and can have no influence whatsoever on the overall
95 result. Among more than 10,000 carefully examined plants there were only a very few in which admixture had doubtlessly occurred. Since such interference was never noticed in the greenhouse, it may be assumed that *Bruchus pisi*, and perhaps also the cited abnormalities in floral structure, are to blame.

Interpretive Comments

Mendel begins this section by pointing out (lines 1–3) that when two plants both showing the same character of a trait are crossed, the character appears in the offspring of the cross unchanged, exactly as it was in the parental plants. If these offspring are allowed to self-fertilize, the character will also appear in their own offspring unchanged.

However, if the characters of a trait are not the same in the two parental plants, these two characters unite to form a new trait in the hybrids produced (lines 4–5). This new trait is not a constant trait. At this point Mendel makes clear to us why he had taken the trouble to set down this preliminary explanation (lines 6–8). He planned to observe the changes that take place when these hybrids, formed from unlike parental characters, reproduce and to try to figure out the law according to which these characters appeared in successive generations.

He then notes that some of the traits he had originally intended to use could not be used because they did not meet criterion number one (section 2, line 9). They did not possess differing traits that were constant. For example, if the trait were flower color, and one parent had red flowers and the other had white flowers, the hybrid might be a little less red than the one parent but not white. The offspring of the hybrids might vary from bright red to pale red to pink to very pale pink to white. These are the sorts of plant characters that Mendel describes as showing differences of "a more or less" sort (line 16). By working only with the clearly differentiated types, he expected to be able to find out whether all the remaining traits he had selected would show the same pattern of behavior when the hybrids reproduced. If they did all show "concordant behavior" he might have data that would enable him "to deduce the law according to which" (line 7) these changes take place.

Mendel then lists, in serial order, the seven traits he intended to use in his experiments, together with the contrasting characters of each. The first two traits are **seed traits;** that is, they are traits visible in the seeds of the fertilized plants. As such they can be observed at the end of the year in which the plant bearing them is fertilized. The remaining traits cannot be observed until the following year, when seeds from the fertilized plant of the preceding year have been planted and grown into a new plant.

There are two other points about the traits Mendel selected that need comment. In describing trait number 2 (lines 28–31) Mendel refers to the color of the seed **albumen**. This term is wrong, because in pea seeds there is *no* albumen. In modern usage, as we pointed out in the section on the botany of peas, the colored structures inside the pea seed coat are called the cotyledons. These structures are the first leaves of the pea embryo. They are engorged with food, mainly starch, which the embryo uses in the process of germination. When the embryo begins to develop, the cotyledons dry up and never rise above the ground. Incidentally, the cotyledons are the hemispheres familiarly known as split peas.

The second point concerns the determination of the color of the cotyledons within the seed. When Mendel, describing the color of the seed coat, uses the term *white* he means transparent. Hence, a seed with a white coat is one with a transparent coat. In this case it is possible to see the cotyledon color through the coat. For the other colors he used, grey or grey-brown and leather brown (line 34), the seed coat is not transparent. When this is the case, the cotyledon color can be determined only if means can be found to open the seed coat without damaging the embryo—for example, by soaking the seed in water, reading the cotyledon color, and then planting the seeds without their seed coat. However, Mendel does not tell us how he determined cotyledon color when the seed coat was opaque.

Both cotyledon color and seed shape are embryonic traits; that is, they belong to the embryo, which is the earliest form of the next generation. These two traits can be observed in the seeds at the end of the growing season in which the flowers of the plant are fertilized.

The remaining seed trait, seed coat color, presents a problem because, despite its name, it is not an embryonic trait as the other two seed traits are. As we noted earlier in the botanical introduction, the seed coat is the part of the maternal plant that surrounds the embryo. It can be compared to the human **amnion** (maternal tissue). When you look at the seed coat you are looking at one generation. When you look at the other two seed traits, cotyledon color and seed shape, you are looking at the *next* generation. The difference in generations of these traits could become a serious problem for a plant breeder or a researcher such as Mendel if not recognized. Even when recognized it could result in real complications when working out the design of plant hybridization experiments. Fortunately, Mendel was well aware of these problems and adapted his methods to take account of them, as we shall see later.

Mendel notes also that certain flower color characters and stem color characters are uniformly associated with the seed coat colors (lines 33–36), but apparently made no use of this concordant behavior. The remaining traits and descriptions are sufficiently clear not to need further comment. Therefore, we shall pass on to the remainder of this section.

Following the listing of traits, Mendel writes out in table form (lines 57–64) the number of fertilizations and the number of plants involved in starting the seven experiments. As you can see, he used 70 plants and made a total of 287 fertilizations. This means that he fertilized about 4 blossoms on each of the 70 plants. The plants used were selected (lines 67–72) from those grown in the two-year field trials that he wrote about in the previous section (section 2, lines 46–57).

Mendel notes (lines 70–72) that he carried out reciprocal fertilizations with each trait and describes the process. The results he obtained here should be noted carefully because later he used the results of these crosses as evidence for some very important conclusions. He tells us that he raised most of the plants for these experiments in garden beds (line 73). We know from other sources that the garden set aside for his use measured about 35 meters long by 7 meters wide. Since a meter is

39.375 inches or 3.28 feet, we can calculate the size of the garden in units more familar to American readers as about 115 feet long and about 23 feet wide. It has been estimated that Mendel could have raised about 5,000 plants per year in this space. We also know from monastery records that he had at least one greenhouse available for his use. The greenhouse would have provided space for the plants used as controls (lines 75–78) against disturbance of results due to fertilization with foreign pollen by insects. As we noted in the introduction to this section, the plants grown in the greenhouse would have given Mendel a good idea of what the offspring of each cross should look like if nothing disturbed the fertilization process. These controls then provided a standard against which the results obtained with plants raised in the outdoor garden could be tested. This use of controls by Mendel was very unusual in biological experiments at this time (about 1855–1863). He probably came into contact with the idea of controls and their uses in his studies of physics and chemistry at the University of Vienna.

The remainder of the material (lines 80–99) is concerned with various ways in which the plants might be contaminated with foreign pollen and so give erroneous results. Recognizing that these deviant forms—"only a very few among 10,000 carefully examined plants"—were the result of foreign pollen, he had no hesitation about leaving them out of his data.

§4

The Form of the Hybrids

In section 3 Mendel wrote about starting his experiments by making a series of artificial fertilizations (crosses). In section 4, he reported on the forms of the hybrids produced. It will be helpful in following his work to have a fairly clear idea of what amounts of time are involved in these kinds of experiments. First of all, Mendel would have planted the peas of the various kinds in the garden when the chance of frost was past, most likely in May. He would then have had to wait until sometime in June for the plants to mature enough to form flowers. At that time he could make his artificial fertilizations. The fertilized blossoms would then have to be protected while the peas formed and ripened. By this time it would probably be September. He could then harvest the seeds, keeping those from each plant in a bag with a tag on it recording what the parent plants were.

With the onset of winter, Mendel could sit down and examine the peas in each bag and record any useful information. We must remember that at this point he could obtain useful information only on seed shape and on seed (cotyledon) color. These two traits are those of the embryo and can be determined by examination as soon as the seeds are ripe. But this is not true of the remaining traits: the seed coat color, the shape of the ripe pod, the color of the unripe pod, the distribution of flowers on the stem, and the difference in stem length. For all of these Mendel had to plant part of the seeds he had carefully collected and grow them to maturity. Some of the traits, the distribution of flowers and the difference in stem length, could be determined before the June flowering, others only later.[1]

Mendel's Text

Experiments on ornamental plants undertaken in previous years had proven that, as a rule, hybrids do not represent the form exactly intermediate between the parental strains. Although the intermediate form of some of the more striking traits, such as those relating to shape and size of leaves, pubescence of individual parts, and so forth, is indeed nearly always seen, in other cases one of the two parental traits is so preponderant that it is difficult, or quite impossible, to detect the other in the hybrid.

The same is true for *Pisum* hybrids. Each of the seven hybrid traits either resembles so closely one of the two parental traits that the other escapes detection, or is so similar to it that no certain distinction can be made. This is of great importance to the definition and classification of the forms in which the offspring of hybrids appear. In the following discussion those traits that pass into hybrid association entirely or almost entirely unchanged, thus themselves representing the traits of the hybrid, are termed *dominating,* and those that become latent in the association, *recessive.* The word "recessive" was chosen because the traits so designated recede or disappear entirely in the hybrids, but reappear unchanged in their progeny, as will be demonstrated later.

All experiments proved further that it is entirely immaterial whether the dominating trait belongs to the seed or pollen parent; the form of the hybrid is identical in both cases. This interesting phenomenon was also emphasized by Gärtner, with the remark that even the most practiced expert is unable to determine from a hybrid which of the two species crossed was the seed plant and which the pollen plant.

Of the differing traits utilized in the experiments the following are dominating:

1. The round or nearly round seed shape with or without shallow depressions.

2. The yellow coloration of seed albumen.

3. The grey, grey-brown, or leather-brown color of the seed coat, associated with violet-red blossoms and reddish spots in the leaf axils.

4. The smoothly arched pod shape.

5. The green coloration of the unripe pod, associated with the same color of stem, leaf veins, and calyx.

6. The distribution of flowers along the stem.

7. The length of the longer stem.

With respect to this last trait it must be noted that the stem of the
40 hybrid is usually longer than the longer of the two parental stems, a fact which is possibly due only to the great luxuriance that develops in all plant parts when stems of very different lengths are crossed. Thus, for instance, in repeated experiments, hybrid combinations of stem 1′ and 6′ long yielded, without exception, stems varying in length from 6′ to 7½′.
45 *Hybrid seed coats* are often more spotted; the spots sometimes coalesce into rather small bluish-purple patches. Spotting frequently appears even when it is absent as a parental trait.

The hybrid forms of *seed shape* and *albumen* develop immediately after artificial fertilization merely through the influence of the foreign
50 pollen. Therefore they can be observed in the first year of experimentation, while the remaining traits do not appear in the plants raised from fertilized seeds until the following year.

Interpretive Comments

Mendel starts this section (lines 1–3) by saying that experiments with ornamental plants have proven that hybrids usually are not intermediate in appearance between the parental plants. But he did not feel that it was necessary to explain why this was important, probably because most of the people in the audience who heard his lectures were well enough trained in biology to know why he said it. The following is the reason. At that time it was commonly accepted that many, if not most, of the characteristics of hybrids *were* intermediate between those of their two parents. This was explained by an analogy to the results obtained when two colored dyes or pigments were mixed. The color that resulted was said to be the result of mixing or blending the original colors into one and other. In similar fashion the parental characters were assumed to blend to give the intermediate hybrid character. Mendel points out (lines 3–8) that this was indeed true of some characteristics but not of all of them. In some instances one of the two parental characters was so strongly expressed that the other was either totally invisible or nearly so. He then states clearly (lines 9–13) that this last example was true in his *Pisum* hybrids. The hybrid character, which was formed by the union of the two contrasting parental characters, was not intermediate between those characters. Instead, the hybrid character was so like one of the parental characters that the other could not be seen at all.

These behaviors offered a useful basis for classifying the characters into two groups. He named the character of one of the parents the *dominating character* and the other the *recessive character* (lines 13–19). It is important that you understand the distinction between the two and know what the names mean, since you will see them very often in Mendel's paper. What Mendel has done here is to take two words from common everyday language and redefine them, giving them new and limited meanings, which he used consistently from this point on. These two terms have thus became "technical terms." This sort of redefinition is a common practice in science, and any attempt to use these terms with their old, familiar meanings unfortunately results in confusion and misunderstanding. Thus, anyone wishing to understand a science must know these special meanings. Note that Mendel chose the word *dominating,* not *dominant,* which some of you may remember from earlier experience with biology. Following Mendel, we shall use the word **dominating** in our discussion. Before Mendel gave his definition of dominating as applied to the behavior of particular characters, it was customary to refer to the most frequently occurring character of a trait as the predominating character. The term also carried the implication that the predominating character was the stronger one, the one that had prevailed in the struggle for expression. However, the character that is the dominating one in Mendel's sense is not always the predominating one in the earlier sense. Therefore, the test of whether a given character is a dominating one is not whether it is the most frequent, but whether or not it conforms to Mendel's definition; that is, whether it is the parental character that appears in the hybrid of two true-breeding parents with contrasting characters.[2]

One last point. If you refer to section 3, lines 19 and 20, you will see that Mendel wanted to find out whether, in hybrid unions, the traits all showed concordant behavior. If one of the characters of each trait always behaves as a dominating character and the other always behaves as a recessive character, he will have found a pattern of behavior common to all the pairs. In this respect they will show a concordant behavior. From this he may be able to derive a generalization or find a clue that will lead him to a law of nature. In fact he has already found one such clue (see lines 20–22 in this section [4]), and you will see (lines 22–25) that Mendel knew what earlier hybridizers had discovered in relation to his problem. In short, he reviewed the literature before he started his experiments. We also know this from his "Introductory Remarks" (section 1, lines 6–13). A literature review is necessary so that the scientist knows what has already been tried and what the results were. If this is not done, a great deal of time may be wasted doing things already done by someone else.

Mendel continues this section by listing the dominating character of each of the seven traits (lines 28–38). Observe that, as we mentioned previously, in trait 2 the yellow color is in the cotyledons (not the albumen). Note also that, in addition to the primary characters that he lists as the traits he will follow in the experiments, he also lists some secondary traits found along with them. These secondary traits are color of the flower, the stalk, the calyx, and the leaf veins and the arrangement of the flowers (the inflorescence). These secondary characters are of interest to botanists but are not the primary differentiating characters that Mendel was following in his experiments.

Mendel's discussion of his results with trait 7 (lines 39–44) merits special comment. The results he observed, that some of the hybrids were taller than the tall parents, are usually referred to as examples of **hybrid vigor.** This same behavior is shown in hybrid corn, the hybrids being taller than the parental lines.

The last paragraph is very important because in it Mendel comments that the seed traits cannot be read at the same time as the other traits. We have already considered the material about the differences in seed traits and plant traits above. Therefore, we can pass on to section 5 of the text.

Notes

1. Mendel never gave us a time schedule of his experiments. To help the reader we have written one for his monohybrid experiments in appendix C.

2. Even though Mendel was very clear in his definition of a dominating character, the term *dominating* was never accepted. After 1900, when Mendel's paper was reintroduced to biologists as a study of inheritance, the word *dominant,* which is a descriptive adjective, was substituted for the term *dominating*. Then, when the inheritance of characters was explained by the inheritance of a pair of genes, the concept of dominance of one character over another was transferred to the genes.

§5

The First Generation from Hybrids

Mendel's task, as he defined it, was to search for a law governing or describing the formation of plant hybrids and the development of their progeny over several generations. In section 3 he described the formation of the plants that were hybrid for the contrasting characters of each of his seven traits. In section 4 he described the forms of those hybrids and noted their concordant behaviors, which became the basis for the first two laws of hybrids:

Law 1. The hybrid offspring of parents, each true-breeding for one of the contrasting characters of a trait, are all alike and like one of the parents. No intermediate types are formed.

Law 2. Reciprocal fertilizations yield the same hybrid forms. That is, the hybrid trait will be that of the dominating parent regardless of whether that is the seed parent or the pollen parent.

Laws such as these, derived from experimental results and not from a theoretical scheme, are known as *empirical laws*. Thus far, Mendel has discovered these two laws but has not called them laws. We have written these two out in our own words, not Mendel's. As he discovers others we will add them to the list. Eventually, as you will see, he will discover one master law that includes all of these earlier ones.

In this section (5) he extends his experiments to the next logical step, the production of the first generation of progeny from those hybrids. To do this he planted the seeds produced by the crosses described in section 3 and waited for them to develop into plants, flower, self-fertilize, and produce seeds. These seeds became the basis for determining a new set of data. Mendel could then examine the data to see whether or not these seeds also showed concordant behaviors and, if they did, whether he could deduce the law appropriate for this stage of the experiment.

As you will discover, he was now confronted with some very large numbers of pea seeds and plants. These large samples created some problems in arranging the data and in determining what conclusions he should draw from them. At the time when Mendel was doing this work, in the mid 1800s, statistical methods of judging whether a sample was of a size adequate to test the quality of results obtained were not as well developed as they are at the present time. We now have reliable methods of testing whether data such as Mendel obtained are too good or too poor to be acceptable. While these methods are not especially difficult, an understanding of them is not strictly necessary here since Mendel did not use them. For those of you who are interested in learning something about them, we have included some examples in appendix D, where the techniques are applied to some of Mendel's data.

Mendel's Text

1 In this generation, *along with the dominating* traits, the *recessive* ones also reappear, their individuality fully revealed, and they do so in the decisively expressed average proportion of 3:1, so that among each four plants of this generation three receive the dominating and one
5 the recessive characteristic. This is true, without exception, of all traits included in the experiments. The angular, wrinkled seed shape, the green coloration of the albumen, the white color of seed coat and flower, the constrictions on the pods, the yellow color of the immature pod, stalk, calyx, and leaf veins, the almost umbellate inflorescence, and the
10 dwarfed stem all reappear in the numerical proportion given, without any essential deviation. *Transitional forms were not observed in any experiment.*

Since the hybrids resulting from reciprocal crosses were of identical appearance and showed no noteworthy deviation in their
15 subsequent development, the results from both crosses may be totaled in each experiment. The numerical proportions obtained for each pair of differing traits are as follows:

Experiment 1. Seed shape. From 253 hybrids 7324 seeds were obtained in the second experimental year. Of them, 5474 were round or roundish and
20 1850 angular wrinkled. This gives a ratio 2.96:1.

Experiment 2. Albumen coloration. 258 plants yielded 8023 seeds, 6022 yellow and 2001 green; their ratio, therefore, is 3.01 : 1.

In these two experiments each pod usually yielded both kinds of seed. In well-developed pods that contained, on the average, six to nine seeds, all seeds were fairly often round (Experiment 1) or all yellow (Experiment 2); on the other hand, no more than 5 angular or 5 green ones were ever observed in one pod. It seems to make no difference whether the pods develop earlier or later in the hybrid, or whether they grow on the main stem, or on an axillary one. In the pods first formed by a small number of plants only a few seeds developed, and these possessed only one of the two traits; in the pods developing later, however, the proportion remained normal. The distribution of traits also varies in individual plants, just as in individual pods. The first ten members of both series of experiments may serve as an illustration:

	Experiment 1 Shape of Seeds		Experiment 2 Coloration of Albumen	
Plant	Round	Angular	Yellow	Green
1	45	12	25	11
2	27	8	32	7
3	24	7	14	5
4	19	10	70	27
5	32	11	24	13
6	26	6	20	6
7	88	24	32	13
8	22	10	44	9
9	28	6	50	14
10	25	7	44	18

Extremes observed in the distribution of the two seed traits in a *single* plant were, in Experiment 1, one instance of 43 round and only 2 angular, another of 14 round and 15 angular seeds. In Experiment 2 there was found an instance of 32 yellow and only 1 green, but also one of 20 yellow and 19 green seeds.

These two experiments are important for the determination of mean ratios, which make it possible to obtain very meaningful averages from a fairly small number of experimental plants. However, in counting the seeds, especially in Experiment 2, some attention is necessary, since in individual seeds of some plants the green coloration of the albumen is less developed, and can be easily overlooked at first. The cause of partial disappearance of the green coloration has no connection with the hybrid character of the plants, since it occurs also in the parental plant; furthermore, this peculiarity is restricted to the individual and not inherited by the offspring. In luxuriant plants this phenomenon was noted quite frequently. Seeds damaged during their development by insects often vary in color and shape; with a little practice in sorting, however, mistakes are easy to avoid. It is almost superfluous to mention that the pods must remain on the plants until they are completely ripe and dry, for only then are the shape and color of the seeds fully developed.

Experiment 3. Color of seed coat. Among 929 plants 705 bore violet-red flowers and grey-brown seed coats; 224 had white flowers and white seed coats; this yields the proportion 3.15 : 1.

Experiment 4. Shape of pods. Of 1181 plants 882 had smoothly arched pods, 299 constricted ones. Hence the ratio 2.95 : 1.

Experiment 5. Coloration of unripe pods. The experimental plants numbered 580, of which 428 had green and 152 yellow pods. Consequently, the former stand to the latter in the portion 2.82 : 1.

Experiment 6. Position of flowers. Among 858 cases, 651 had axillary flowers and 207 terminal ones. Consequently, the ratio is 3.14 : 1.

Experiment 7. Length of stem. Of 1064 plants, 787 had the long stems, 277 the short stems. Hence a relative proportion of 2.84 : 1. In this experiment the dwarfed plants were tenderly lifted and transferred to beds of their own. This precaution was necessary because they would have become stunted growing amidst their tall brothers and sisters. Even in the earliest stages they can be distinguished by their compact growth and thick dark-green leaves.

When the results of all experiments are summarized, the average ratio between the number of forms with the dominating trait and those with the recessive one is 2.98 : 1, or 3 : 1.

The dominating trait can have *double significance* here—namely that of the parental characteristic or that of the hybrid trait. In which of the two meanings it appears in each individual case only the following generation can decide. As parental trait it would pass unchanged to all of the offspring; as hybrid trait, on the other hand, it would exhibit the same behavior as it did in the first generation.

Interpretive Comments

In the preceding section, "The Form of the Hybrids," Mendel demonstrated that the character shown by the hybrids in each of the seven experiments resembled one of the parental characters so closely that the other parental character could not be detected. The character shown by the hybrid in each case was the dominating one. In this section he points out that the hybrids did not breed true but were variable, with both parental characters reappearing in the progeny. He reports on experiments in which he can observe these changes in each of the hybrid traits. His aim in doing these experiments is to try to "deduce the law according to which [these changes] appear in successive generations" (section 3, lines 7 and 8).

He begins section 5 with a generalization about what the changes were (lines 1–3) and states the numerical relationship of the two classes of offspring to one another: three of the dominating character to one of the recessive character for each of the seven traits. Thus, when each of the hybrid parents was allowed to self-fertilize, the progeny were not all like the parent, which would be the case if these parents were true-breeding. Part of the progeny, ¾ of them, showed the dominating character and appeared to be like the dominant parent of the hybrid and like that of the hybrid itself. Part of the progeny, ¼ of them, showed the recessive character, like that of the recessive parent of the hybrid. These characteristics reappeared intact. Mendel lists (lines 6–11) the actual recessive characters that could be identified. Here he reaffirms that no transitional forms were observed (lines 11–12). In stating this he implies that no blending of parental characters has taken place, but he does not say it.

He comments (lines 13–16) that the hybrids of each trait formed by reciprocal fertilization showed essentially the same appearance and behaved like the others when self-fertilized. Therefore, the numerical data obtained from them could be added to that of the others to give the totals, which he reported in subsequent pages of his papers.

He gives the data for each of the two seed traits, seed shape and albumen coloration (cotyledon color) on lines 18–22 and gives the summary ratios of the contrasting characters, 2.96 to 1 and 3.01 to 1. As you can see, each of the experiments involved many plants and yielded many seeds. On each plant the fertilization of a single flower resulted in a single pod. Mendel gives some data (lines 23–32) about how the number and kinds of peas varied from pod to pod and how a variety of factors affected these numbers.

In addition to variations within individual plants from one pod to another there were variations between plants. To illustrate this, Mendel gives the data for the distribution of the contrasting characters for 10 plants of Experiments 1 and 2 (lines 35–47). In experiment 1 there were 336 round to 101 angular peas for an average of 3.33 to 1 based on totals. If we look at the ratios of round to angular for each plant, they range from 1.90 to 1 for plant 4 to 4.67 to 1 for plant 9. In experiment 2 there were 355 yellow cotyledon to 123 green cotyledon peas for an average ratio of 2.89 to 1. The ratios for the individual plants range from 1.85 to 1 for plant 5 to 4.89 to 1 for plant 8. Thus, it is obvious that smaller samples (progeny from one plant) are not as indicative of the "true" or ideal ratios as larger samples (progeny of many plants), In order to obtain reliable counts, such as those Mendel reports, certain precautions had be taken to avoid errors in classification of the seeds. Mendel reviews these in lines 55–68. Although these would be important if we were doing the experiments, we can pass over them lightly and examine the data for experiments 3 through 7. As you can see, the ratios all cluster around or approach 3 to 1 as the limiting, or true, value as Mendel indicates (lines 86–88).

It is important to pause here and look at the process Mendel used in reducing the chaos of all these varying ratios from the different plants to the whole number 3:1 ratio. One reason why it is important to understand this is because Mendel will use the same method or technique again and again in the rest of his experiments. A second reason is that he was the first person, as far as we know, to apply this method in biology. No one before him had thought about hybridization in this quantitative way or sought quantitative laws to describe the process. The third reason is that it gives an example of the way Mendel adapted ideas and methods of the physical sciences to uses in the biological sciences. This was one of the benefits of Mendel's interdisciplinary education.

Here he adapted two mathematical techniques and a chemical concept to solve a biological problem. The first mathematical technique is that the arithmetic mean is the best single representation of a group of measurements. The second mathematical technique is that, given a series of measurements of a quantity whose value is influenced by chance errors, the larger the sample of measurements the closer the value comes to its true or ideal value. The chemical concept that Mendel adapted to hybridization is the law of constant composition. This law says, in effect, that in a pure chemical compound the component elements are always present in a constant proportion by weight. For example, when a sample of pure water is decomposed by a direct electric current, it always produces oxygen and hydrogen in the weight ratio of 8 to 1. Each such determination is subject to various experimental errors. Some ratios will be greater than 8 to 1, some less than 8 to 1. An overall average of these comes closer to 8 to 1 as the number of determinations increases until it finally becomes clear that the best value is the 8 to 1 ratio.

Mendel adapted this mathematical-chemical concept to studying the "composition" of hybrids when they split up into their components in the formation of offspring. He was able to show, as we have just seen, that the offspring of self-fertilized monohybrids always consisted of three of the dominating-form progeny to each one of the recessive form. He discovered this ratio in an interesting way. He recognized, first, that one pea plant does not produce enough seeds to give a good sample and, second, that the number of seeds produced by one plant is subject to a large number of chance factors, as shown in the example of ten plants that he gave earlier. To overcome this inadequacy of data from a single plant he recorded the numbers of each kind of offspring for each plant, then added the numbers of each kind together, and then calculated their ratios. He then went one step further and computed the average of the seven ratios, which was the most representative ratio for all the experiments. This, when rounded to the nearest whole number, gave him the most probable and ideal ratio of 3:1.

In the remaining lines (89–94) Mendel draws attention to the fact that while the parental dominating trait and the hybrid dominating trait had the same appearance, they did not behave in the same way. The parental dominating trait is true-breeding, as Mendel had shown in his two-year field trials. Whether this was also true for the dominating trait produced from the hybrids could not be determined by visual examination. It could only be determined by examining the characteristics of the next generation produced by self-fertilization.

Let us take the shape of the seed as an example of a trait. According to Mendel, the hybrid between a true-breeding round-seeded variety and a true-breeding angular-seeded variety will be round. Among the seeds produced by the hybrid there will be three round seeds to one angular. The question is, will these round seeds give rise to plants true-breeding for round seeds or will some of them produce angular as well as round seeds? There is no way to know except by planting the round seeds individually, growing the plants from them, letting the flowers self-pollinate, harvesting the seeds, and checking them for shape.

Mendel carefully distinguishes the differences to be expected in each case (lines 89–94). These comments set the stage for the next section of his report, in which he proceeded to find out how the hybrids of each trait do behave.

§6

The Second Generation from Hybrids

We have seen how Mendel's quantitative studies on the behaviors of monohybrids revealed some patterns of concordant behaviors. These became the basis for the first two empirical laws of the formation and development of hybrids. The results of the experiments in section 5 enabled him to discover a third law, which we have added below (again given in our own words, not his).

Law 1. The hybrid offspring of parents, each true-breeding for one of the contrasting characters of a trait, are all alike and like one of the parents. No intermediate types are formed.

Law 2. Reciprocal fertilizations yield the same hybrid forms. That is, the hybrid trait will be that of the dominating parent regardless of whether that is the seed parent or the pollen parent.

Law 3. When the hybrids are allowed to self-fertilize, the offspring always appear in two classes: one class like the hybrids and like one of the original true-breeding parents (the dominating); and one class like the parental character not visible in the hybrid generation (the recessive). No intermediate forms are produced. The two classes occur in approximate ratio of 3 dominating to 1 recessive.

This discovery led Mendel to ask whether there were other such laws to be found if he examined the second generation from the hybrids. One problem to be investigated was how the dominating fraction of the progeny of the first generation from the hybrids would behave when used to produce new progeny. The second problem to be investigated was the behavior of the recessive fraction of the progeny from the first generation. As you already know, Mendel had some ideas about what might be true. However, the only way to really find out was to plant the seeds of the first generation from the hybrids and wait. Would the plant growing from the seed be the same as the parental plant that bore the seed? There are only two possible answers—yes or no. If the answer is yes, then the parental plant was true-breeding. If the answer is no, then the parental plant was not true-breeding, but was hybrid. Mendel used this method with *every seed* he planted, thousands of them. This meant that he could and did trace the genealogy of every seed, from plant to seed to plant to seed, sometimes for as many as six generations. Mendel was a very thorough scientist. Not only was he a skilled experimenter, but he was also a meticulous record keeper—and as we shall see, a very careful and insightful analyzer of data.

Mendel's Text

1 Those forms that receive the recessive character in the first generation do not vary further in the second with respect to this trait; they remain *constant* in their progeny.

The situation is different for those possessing the dominating trait
5 in the first generation. Of these, *two* parts yield offspring that carry the dominating and the recessive trait in the proportion of 3:1, thus showing exactly the same behavior as the hybrid forms; only *one* part remains constant for the dominating trait.

The individual experiments yielded results as follows:

10 *Experiment 1*. Among 565 plants raised from round seeds of the first generation, 193 yielded only round seeds, and therefore remained constant in this trait; 372, however, produced both round and angular seeds in the proportion of 3:1. Therefore, the number of hybrids compared to that of the constant breeding forms is as 1.93:1.

15 *Experiment 2*. Of 519 plants raised from seeds whose albumen had yellow coloration in the first generation, 166 yielded exclusively yellow, while 353 yielded yellow and green seeds in the proportion 3:1. Therefore, a partition into hybrid and constant forms resulted in the proportion 2.13:1.

20 For each of the subsequent experiments, 100 plants that possessed the dominating trait in the first generation were selected, and in order to test this trait's significance 10 seeds from each plant were sown.

Experiment 3. The offspring of 36 plants yielded exclusively grey-brown seed coats; from 64 plants came some grey-brown and some white coats.

25 *Experiment 4*. The offspring of 29 plants had only smoothly arched

pods; of 71, on the other hand, some had smoothly arched and some constricted ones.

Experiment 5. The offspring of 40 plants had only green pods; those of 60 plants had some green, some yellow.

Experiment 6. The offspring of 33 plants had only axillary flowers; of another 67, on the other hand, some had axillary, some terminal flowers.

Experiment 7. The offspring of 28 plants received the long stem, those of 72 plants partly the long, partly the short one.

In each of these experiments a certain number of plants with the dominating trait become constant. In evaluating the proportion in which forms with the constantly persisting trait segregate, the first two experiments are of special importance, since in these a fairly large number of plants could be compared. The ratios 1.93:1 and 2.13:1 taken together give almost exactly the average ratio of 2:1. Experiment 6 gives an entirely concordant result; in other experiments the proportion varies more or less, as was to be expected with the small number of 100 experimental plants. Experiment 5, which showed the greatest deviation, was repeated, and instead of 60:40, the ratio of 65:35 was obtained. *Accordingly, the average ratio of 2:1 seems ensured.* It is thus proven that of those forms which possess the dominating trait in the first generation, two parts carry the hybrid trait, but the one part with the dominating trait remains constant.

The ratio of 3:1 in which the distribution of the dominating and recessive traits takes place in the first generation therefore resolves itself into the ratio of 2:1:1 in all experiments if one differentiates between the meaning of the dominating trait as a hybrid trait and as a parental character. Since the members of the first generation originate directly from the seeds of the hybrids, *it now becomes apparent that of the seeds formed by the hybrids with one pair of differing traits, one half again develop the hybrid form while the other half yield plants that remain constant and receive the dominating and the recessive character in equal shares.*

Interpretive Comments

In the concluding paragraph of section 5, "The First Generation from Hybrids," Mendel pointed out that the dominating character shown by three-fourths of that first generation might be either the parental dominating character, which is true-breeding, or it may be the hybrid dominating character, which is not true-breeding. The only way to find out which of these is present in each individual case (plant or seed) is to raise the next generation. This he proceeded to do, treating each trait separately.

In the first two paragraphs of section 6 (lines 1–9) Mendel gives a brief summary of the results he found when he performed the experiments he wrote about at the end of section 5. There he wrote that plants showing the recessive character of a trait bred true, while those showing the dominating character of each trait did not always do so. In fact, only one plant out of the three showing the dominating character bred true; the remaining two plants out of those three did not. Those two produced offspring in the same 3 to 1 ratio as the original hybrids. Therefore, they were hybrid. Thus, Mendel resolved the original 3 to 1 ratio into a 1 to 2 to 1 ratio in which one-fourth of the total showed the dominating character in true-breeding form, one-half showed the dominating character in the hybrid form, and one-fourth showed the recessive character in true-breeding form. The rest of this section consists of reports of the results obtained in the original experiments.

In this series of experiments Mendel gave special emphasis to the first two, because in those he could obtain his results at the end of the first year and because in them he had a larger body of data from which to reason. In the remaining experiments in this section he used smaller numbers of plants in order to save space in the experimental garden. This was necessary because he was carrying out other experiments with peas at the same time as he was doing the experiments described in this section.

In the first experiment he planted round seeds, the dominating form, from the experiment on the first generation from the hybrids. He obtained 565 plants from these seeds. He found that 193 yielded only round seeds and thus were true-breeding, or as he says, "remained constant in this trait" (lines 11–12). The remaining 372 plants produced both round and angular seeds (sometimes in the same pod) in the same 3 to 1 ratio as the original hybrids had. The ratio of plants producing both kinds of seeds (the hybrids) to plants producing only one kind of seed (true-breeding dominating types) was therefore 372 to 193 or 1.93 to 1. As you can see, this is very close to the 2 to 1 ratio noted above.

In experiment 2 the end results were very similar: 353 plants produced both yellow and green seeds, while 166 plants produced only yellow seeds. In this experiment the ratio of hybrid plants (producing yellow and green seeds) to the true-breeding plants (producing only yellow seeds) was 353 to 166 or 2.13 to 1. Again the ratio was very nearly 2 to 1.

The remaining experiments of the set required a somewhat different procedure, since, unlike the two seed traits just considered, they could not be observed in the same year in which a cross was made. Instead they had to be observed in the following year for the reason already discussed. In these experiments, as noted above, Mendel did not use all the seeds available from the first generation of hybrids. Instead, he used a sample drawn from those plants showing the dominating traits. He selected 100 of these plants for each trait and then selected 10 seeds from each plant for planting in the garden. Each group of 10 seeds was kept separate. He thus had 100 plots of 10 seeds each.

In the third experiment, he found that 36 of these plots yielded plants that, at the end of the season, had only grey-brown seed coats. Each of these plots corresponded to one of the original 100 plants. He therefore reasoned that 36 of the 100 plants were true-breeding for the dominating seed coat character. The remaining 64 plots showed both grey-brown and white seed coats. From this he concluded that 64 of the original plants were not true-breeding, but were hybrid. The ratio of hybrid plants to true-breeding plants was therefore 64 to 36 or 1.78 to 1. Again the ratio was nearly 2 to 1, but not as close as in traits 1 and 2.

In experiments 4 through 7 Mendel used this same sampling procedure and calculation. We shall consider only the ratios he obtained and their meaning. In experiment 4 the ratio of hybrid dominating plants to true-breeding dominating plants was 71 to 29, or 2.45 to 1. In experiment 5 the ratio Mendel obtained was 60 to 40, or 1.50 to 1. Since this ratio was quite far from 2 to 1, he decided to repeat the experiment as a check on the data. When he did so he obtained a ratio of 65 to 35, or 1.85 to 1, which he considered to be acceptable (see appendix D). Such variations in the data are probably attributable, at least in part, to the small size of the sample he used (10 seeds). Since his time, studies of sampling have shown that in Mendel's experiments a sample of about 20 seeds would have been needed to give more reliable results. In experiment 6 he found that the ratio was 67 to 33, or 2.03 to 1, and in experiment 7 it was 72 to 28, or 2.57 to 1. The grand average of the seven experiments was 2.06 to 1, which was close enough for Mendel to comment (line 44) that the 2 to 1 average ratio "seems ensured."

The real significance of this ratio is fully explained in the remainder of section 6, where he points out that the 3 : 1 ratio of the first generation has been resolved into the 1 to 2 to 1 ratio in this second generation from the hybrids. He has shown that the hybrids did not breed true but split up in such a way that not only have both parental lines been recovered in true-breeding form, but the hybrid forms have reappeared as half of the total number of offspring. Here again, as in earlier sections, no forms intermediate or transitional between the parental forms have appeared. Having followed the development of hybrids in their progeny through two successive generations, Mendel proposed to consider what the results would be in subsequent generations if development always followed the same pattern. This is the subject of the next section, in which Mendel introduces some startling ideas.

§7

The Subsequent Generations from Hybrids

Having produced the hybrids for all seven traits, examined the progeny of the first and second generations from these hybrids, and arrived at several empirical laws, in section 7 Mendel extends his study to subsequent generations of progeny from the hybrids.

This is a particularly important section because in it Mendel introduces two new conceptions. One of these is his system of abstract symbolic notation, which greatly simplified writing down the progeny of crosses and also simplified thinking about them. In it the characters present in each type of offspring are represented, as well as whether a given character is dominating or recessive, true-breeding or hybrid. The value of this system, which, with modification, became the foundation of the system still used in genetics, will become evident in the next section of the paper when we follow his two-trait and three-trait hybrid crosses.

The other new conception involves combining the law of development for single-trait hybrids, which he has just discovered, with some simple mathematics to provide a quantitative explanation for a long-known phenomenon, the tendency of hybrids to revert to their parental types when reproduced by self-fertilization for several generations.

Mendel's Text

1 The proportions in which the descendants of hybrids develop and split up in the first and second generations are probably valid for all further progeny. Experiments 1 and 2 have by now been carried through six generations, 3 and 7 through five, and 4,5 and 6 through four without any

deviation becoming apparent, although from the third generation on a small number of plants were used. In each generation the offspring of the hybrids split up into hybrid and constant forms according to the ratios 2:1:1.

If *A* denotes one of the two constant traits, for example, the dominating one, *a* the recessive, and *Aa* the hybrid form in which both are united, then the expression

$$A + 2Aa + a$$

gives the series for the progeny of plants hybrid in a pair of differing traits.

The observation made by Gärtner, Kölreuter, and others, that hybrids have a tendency to revert to the parental forms, is also confirmed by the experiments discussed. It can be shown that the numbers of hybrids derived from one fertilization decrease significantly from generation to generation as compared to the number of newly constant forms and their progeny, yet they can never disappear entirely. If one assumes, on the average, equal fertility for all plants in all generations, and if one considers, furthermore, that half of the seeds that each hybrid produces yield hybrids again while in the other half the two traits become constant in equal proportions, then the numerical relationships for the progeny in each generation follow from the tabulation below, where *A* and *a* again denote the two parental traits and *Aa* the hybrid form. For brevity's sake one may assume that in each generation each plant supplied only four seeds.

Generation	*A*	*Aa*	*a*	Expressed in terms of ratios *A* : *Aa* : a
1	1	2	1	1 : 2 : 1
2	6	4	6	3 : 2 : 3
3	28	8	28	7 : 2 : 7
4	120	16	120	15 : 2 : 15
5	496	32	496	31 : 2 : 31
n				$2^n - 1 : 2 : 2^n - 1$

In the tenth generation, for example, 2^n-1 = 1023. Therefore, of each 2048 plants arising in this generation, there are 1023 with the constant dominating trait, 1023 with the recessive one, and only 2 hybrids.

Interpretive Comments

To recap the results of his experiments, thus far we can say that Mendel has developed the following laws:

Law 1. The hybrid offspring of parents, each true-breeding for one of the contrasting characters of a trait, are all alike and like one of the parents. No intermediate types are formed.

Law 2. Reciprocal fertilizations yield the same hybrid forms. That is, the hybrid trait will be that of the dominating parent regardless of whether that is the seed parent or the pollen parent.

Law 3. When the hybrids are allowed to self-fertilize, the offspring always appear in two classes: one class like the hybrids and like one of the original true-breeding parents (the dominating); and one class like the parental character not visible in the hybrid generation (the recessive). No intermediate forms are produced. The two classes occur in approximate ratio of 3 dominating to 1 recessive.

Law 4. (a) When the recessive offspring of the hybrids are allowed to self-fertilize, they always breed true. (b) When the dominating offspring of the hybrids are allowed to self-fertilize, approximately one-third of them breed true while two-thirds of them behave exactly like the hybrid generation.

It was this pattern that Mendel believed would hold for all subsequent generations. It was this 1 to 2 to 1 pattern that Mendel represented with the abstract symbolic notation $A + 2Aa + a$.

We refer to this notation as abstract because it represents a pattern common to all seven traits he studied, not just one, and because it could be extended to other traits that he did not study. We refer to it as symbolic because, instead of writing out the names of the characters involved, Mendel chose to use letters. Thus, for any of the seven traits, the symbol A represents the true-breeding dominating character, the symbol a represents the true-breeding recessive character, and the symbol Aa represents the dominating hybrid character. By extension, the same symbols and relationships apply to the contrasting character of any trait.

Mendel combined these symbols to give what he called the **developmental series** or the **combination series** or the **empirical simple**

series for the progeny of plants hybrid in a pair of contrasting characters of a trait. That series was, of course,

$$A + 2Aa + a$$

Mendel used all of the different names for this series of symbols that we have indicated above. He referred to it as a "developmental series" because it showed the series of forms in which the progeny of a single-trait hybrid developed. He used the term "combination series" because the series represents the possible combinations of these characters resulting from the self-fertilization of a one-trait hybrid. He referred to the series as an "empirical series" because he developed it to express the results of concordant behaviors in his seven traits, rather than deducing it from a theory about hybrids that he had in mind before starting his experiments. We shall see in the next section of his report why he refers to the series as "simple."

Mendel next turned to a discussion of the phenomenon of **reversion**, an important issue in Mendel's time. Reversion is the return of the hybrid form to the parental forms.[1] Such forms were sometimes spoken of as "throwbacks." Here Mendel demonstrated how his series formula, together with two additional assumptions, enabled him to explain quantitatively the pattern of reversion in single-trait hybrids. Out of this demonstration came the unexpected conclusion that although the relative number of hybrids decreases from generation to generation, they will not disappear from the population of progeny.

In addition, the generalization expressed in the last line of the right half of the table (line 35) enabled Mendel to calculate how many true-breeding dominating forms, how many true-breeding recessive forms, and how many hybrid forms there would be in any generation. This gave the total size of the population of progeny and the distribution of its members among the three types.

You can see how he derived the $2^n - 1$ portion of the expression if you look at the *A* column of the tabulation below:

Generations	*n*	*A*
1	$2 = 2^1$	$1 = 2 - 1 = 2^1 - 1$
2	$4 = 2^2$	$3 = 4 - 1 = 2^2 - 1$
3	$8 = 2^3$	$7 = 8 - 1 = 2^3 - 1$
4	$16 = 2^4$	$15 = 16 - 1 = 2^4 - 1$
5	$32 = 2^5$	$31 = 32 - 1 = 2^5 - 1$
n		$2^n - 1$

Of course, the formula for *n* generations is good only given the stated assumptions. If this seems a strange way for a plant breeder to reason, remember, Mendel was a very unusual breeder—he had been trained in physics and mathematics. In fact, Mendel was the first person to explore the mathematical consequences of self-fertilization in a special population in this way. In this Mendel was a true pioneer.

Notes

1. Reversion has often been observed when the hybrids are grown close to the parental varieties. It can be obtained easily by pollinating the hybrid with the pollen of one of the parents through successive generations. Darwin was interested in reversion. He said: "As a general rule crossed offspring in the first generation are nearly intermediate between the parents, but the grandchildren and succeeding generations continually revert, in a greater or lesser degree, to one or both progenitors" (*The Variation of Animals and Plants under Domestication*, 2d ed. New York, 1900, 22).

§8

The Offspring of Hybrids in Which Several Differing Traits Are Associated

Up to this point Mendel had been paying close attention to the behavior of hybrids formed from contrasting characters of individual traits. In each trait one of the characters used to form the hybrid was dominating, the other recessive. In these experiments the hybrid character always looked the same as the dominating parental character. In the series of experiments we are about to consider, one of the things Mendel investigates is whether this generalization is also true for hybrids in which two, three, or more traits are being studied and in each of which one character is dominating, the other recessive. He finds, as you will see, that the generalization held in these hybrids also, giving him another example of concordant, or lawful, behavior.

Although this is an interesting discovery it is somewhat off the main line of Mendel's investigation, the search for a "generally applicable law of the formation and development of hybrids." Thus far he had discovered four quantitative empirical laws, which have been summarized above. At this point it will simplify our communication if we introduce three new terms, which Mendel did not use but which are now common in studies of breeding and inheritance. What we have called single-trait hybrids are now called **monohybrids;** two-trait hybrids are called **dihybrids;** and three-trait hybrids are called **trihybrids.** Hybrids of many traits are called **polyhybrids.** From here on, we shall use these terms in referring to the various kinds of hybrids.

The master law for the monohybrids states that the progeny of monohybrids split up into a series of forms such that one-fourth are true-breeding for the dominating character (A), one-fourth are true-breeding for the recessive character (a) and one-half are again hybrids (Aa). The series of combinations of characters is represented by the simple developmental series $A + 2Aa + a$.

Having found this law of development for the progeny of monohybrids, Mendel decided that his next task was to find out whether this law would be changed in any way if the first, or *A* trait was combined with a second one in a dihybrid. To study this, Mendel crossed two parental lines, which differed in seed shape and seed color, both of which he had studied separately. The answer to this question of whether the law would change or stay the same could not be found until he had produced the progeny of the second generation from the hybrids, for this was the point at which the evidence became clear with the monohybrids.

The series of progeny for the second generation from the hybrid was, of course, much longer and more complex for the dihybrid than it was for the monohybrid. The monohybrid series consisted of three different groupings of characters *A, Aa,* and *a* in the proportions of 1:2:1. The corresponding dihybrid series consisted of nine different groupings of characters. To establish that the law governing the development of monohybrids had not been changed by introducing the second trait, Mendel had to be able to find the same grouping of characters, $A + 2Aa + a$, given above for the individual traits in those same proportions as part of the longer and more complex dihybrid series.

In the course of searching for this sequence for the *A* trait, he not only found that it was present unchanged but also found, as would be expected, the same sequence for the second trait used to form the dihybrid. Furthermore, Mendel made the astonishing discovery that the complex series of traits from the dihybrid is really a combination of the two simple series and, in fact, can be produced mathematically by means of a term-by-term combination of the simple developmental series for the individual traits. This led him to the discovery of his combination series law for the development of the progeny of hybrids, the "generally applicable law" for which he had been searching.

In addition, his discovery that the combination series for dihybrids could be generated by a term-by-term combination of the simple developmental series for two traits gave him a means of predicting what the series should be for any number of traits he had selected. As you will see, he was able to test both the combination series law and this method of generating a series with his experiments on trihybrids. His experiences with the di-hybrids and trihybrids led him to yet another major generalization, or empirical natural law, about the behaviors of the individual pairs in these hybrid associations.

By now it is probably evident to you that each succeeding section of

Mendel's paper has become somewhat longer and a bit more difficult to follow. However, each of them has dealt with a single experimental problem. So long as this was true, it seemed to us appropriate to leave the text of each section undivided and to place the interpretive comments after the text. However, from this section on, Mendel deals with more than one experimental problem in each section and the reporting and interpretation of each section is more abstract and difficult to grasp.

To make it easier for you to understand what Mendel is reporting and explaining, we have divided the text of each succeeding section into two or more parts, each part pertinent to only one experimental problem or one main idea. We have placed the interpretive text relating to each block of Mendel's text immediately after it. This is followed by an introduction to the next block from the section then the text, and finally its interpretive text. In this section, block A is concerned with the dihybrid experiments and block B with the trihybrid experiments.

Mendel's Text
Block A: The Dihybrid Experiments

1 In the experiments discussed above, plants were used which differed
in only one essential trait. The next task consisted in investigating whether
the law of development thus found would also apply to a pair of differing
traits when several different characteristics are united in the hybrid
5 through fertilization.

The experiments demonstrated throughout that in such a case the
hybrids always resemble more closely that one of the two parental plants
which possesses the greater number of dominating traits. If, for
instance, the seed plant has a short stem, terminal white flowers, and
10 smoothly arched pods, and the pollen plant has a long stem, violet-red
lateral flowers, and constricted pods, then the hybrid reminds one of the
seed plant only in pod shape, and the remaining traits resemble those of
the pollen plant. Should one of the two parental types possess only
dominating traits, then the hybrid is hardly or not at all
15 distinguishable from it.

Two experiments were carried out with a larger number of plants. In
the first experiment the parental plants differed in seed shape and
coloration of albumen; in the second in seed shape, coloration of
albumen, and color of seed coat. Experiments with seed traits lead most
20 easily and assuredly to success.

To simplify a survey of the data, the differing traits of the seed plant will be indicated in these experiments by *A, B, C,* those of the pollen plant by *a, b, c,* and the hybrid forms of these traits by *Aa, Bb,* and *Cc.*

First experiment:

AB seed plant,	*ab* pollen plant,
A shape round,	*a* shape angular,
B albumen yellow,	*b* albumen green.

The fertilized seeds were round and yellow, resembling those of the seed plant. The plants raised from them yielded seeds of four kinds, frequently lying together in one pod. From 15 plants a total of 556 seeds were obtained, and of these there were:

315 round and yellow,
101 angular and yellow,
108 round and green,
32 angular and green.

All were planted in the following year. Eleven of the round yellow seeds did not germinate and three plants from such seeds did not attain fruition. Of the remaining plants:

38 had round yellow seeds … *AB*
65 had round yellow and green seeds … *ABb*
60 had round yellow and angular yellow seeds … *AaB*
138 had round yellow and green and angular yellow and green seeds … *AaBb*

Of plants grown from angular yellow seeds, 96 bore fruit; of these

28 had only angular yellow seeds … *aB*
68 had only angular yellow and green seeds … *aBb*

Of plants grown from 108 round green seeds, 102 bore fruit; of these

35 had only round green seeds … *Ab*
67 had round and angular green seeds … *Aab*

The angular green seeds yielded 30 plants with identical seeds throughout; they remained constant . . . . *ab*

Thus the offspring of hybrids appear in nine different forms, some of them in very unequal numbers. When these are summarized and arranged in order, one obtains:

38	plants	with	the	designation	*Ab*
35	"	"	"	"	*Ab*
28	"	"	"	"	*aB*
30	"	"	"	"	*ab*
65	"	"	"	"	*ABb*
68	"	"	"	"	*aBb*
60	"	"	"	"	*AaB*
67	"	"	"	"	*Aab*
138	"	"	"	"	*AaBb*

All forms can be classified into three essentially different groups. The first comprises those with designations *AB, Ab, aB, ab;* they possess only constant traits and do not change any more in following generations. Each one of these forms is represented 33 times on the average. The second group contains the forms *ABb, aBb, AaB, Aab;* these are constant for one trait, hybrid for the other, and in the next generation vary only with respect to the hybrid trait. Each of them appears 65 times on the average. The form *AaBb* occurs 138 times, is hybrid for both traits, and behaves exactly like the hybrid from which it is descended.

Comparing the numbers in which the forms of these groups occur, one cannot fail to recognize the average proportions of 1:2:4. The numbers 33, 65, 138 give quite satisfactory approximations to the numerical proportions 33, 66, 132.

Accordingly, the series consists of nine terms. Four of them occur once each and are constant for both traits; the forms *AB, ab* resemble the parental types, the other two represent the other possible constant combinations between the associated traits *A, a, B, b.* Four terms occur twice each and are constant for one trait, hybrid for the other. One term appears four times and is hybrid for both traits. When, therefore, two kinds of differing traits are combined in hybrids, the progeny develop according to the expression:

$$AB + Ab + aB + ab + 2ABb + 2aBb + 2AaB + 2Aab + 4AaBb.$$

Indisputably, this series is a combination series in which the two series for the traits *A* and *a*, *B* and *b* are combined term by term. All the terms of the series are obtained through a combination of the expressions:

90 $A + 2Aa + a$
$B + 2Bb + b$

Interpretive Comments

Mendel's opening sentence (lines 1 and 2) gives us very interesting and important information about the experiments he had just completed. He wrote there that the plants he had used so far in these monohybrid experiments differed "in only one essential trait." There is a tendency to think that the plants differed *only* in the characters of the one trait whose behavior he was studying and that they bred true in all other traits they showed. In fact, the first statement in section 3 (lines 1–5) would support that idea. However, lines 1 and 2 of the present section suggest that the plants he was working with may have differed in several traits, while he was following the behavior of the two characters of only one trait and ignoring whatever was happening to the other traits. We shall see that this idea of how he had been working is further supported by the illustration he gives in lines 6 through 15. Thus, although Mendel was usually working with polyhybrids, in each of the first seven sections he was following the behavior of the characters of only one trait. However, in this section, he is at first following the behavior of the characters of two traits, then, later, three traits.

Before going any further we should go back to the statements of lines 6 through 15. To make Mendel's example easier to understand we have rearranged the information:

	Seed plant	Pollen plant
1.	*R* stem short	*D* stem long
2.	*R* terminal flowers	*D* lateral flowers
3.	*R* white flowers	*D* violet red flowers
4.	*D* smoothly arched pods	*R* constricted pods

In the left-hand column we have entered the information he gave us about the seed plant, and in the right-hand one the information about the pollen plant. To the left of each character of each plant we have indicated whether the character is dominating (*D*) or recessive (*R*). You can always check these by going back to section 4 to see how Mendel listed them.

Now suppose the two parental plants are crossed. What will the polyhybrid look like? We can see that it should be long-stemmed, with violet red flowers distributed along the stem, and the pods should be smoothly arched, because those characteristics are the dominating ones. Thus far the hybrid always shows the dominating characteristic no matter which parent has it. Hence, the hybrid should look more like the parent with the most dominating characteristics rather than the one with the most recessive characteristics. Mendel explained this comparison in appearance of the hybrid and the parental plants (lines 11–13).

We can now turn to the first of the two experiments in this section, the dihybrids. Note that at first Mendel used both verbal and letter symbols in describing these experiments (lines 21–27). Later, the verbal descriptions will disappear entirely.

In line 28 Mendel states that the seeds produced by crossing the two parents were all round and yellow like the seeds of the seed parent. The corresponding characters of the pollen plant, being recessive, were not visible within these seeds. This agrees with the expectation derived from his monohybrid experiments. Mendel does not tell us how many seeds were produced nor whether he planted all of them the next spring. We do know that he harvested 556 seeds from 15 plants. Probably, there were more plants than this and Mendel selected 15 of the healthiest and most vigorous to serve as seed producers for the next step in the experiments. When the seeds were harvested, classified by appearance, and counted, he had the results shown in lines 32 through 35.

These seeds corresponded to those obtained from the monohybrids in the first generation from the hybrids. Those were always in very nearly a 3 to 1 ratio of dominating to recessive characters. Remember that in these experiments Mendel had to raise the second generation from the hybrids to find out which of the plants were hybrids and which were true-breeding. If a plant produced only one type of seeds, all round or all angular, that plant was true-breeding for that character. If, on the other hand, a plant gave rise to two kinds of seed, for example, both round and angular, that plant must have been hybrid. In this way

Mendel was able to determine the breeding structure of each of the plants. He followed the same procedure with the dihybrids. But here the judgment has to be made in each plant for each of the two traits involved. Thus, beginning with the second generation from the hybrid, seeds produced by each plant were kept together and separate from those produced by the other plants.

In lines 39 through 49 Mendel has recorded the visible characters of the seeds produced by each plant. The first group (lines (39–43) were all raised from the yellow round seeds of the first generation from the hybrids. Thirty-eight of these when self-fertilized produced only round and yellow seeds. Therefore, these plants were true-breeding for both traits. Since round is the dominating character of the seed-shape trait and yellow is the dominating character of the seed-color trait, Mendel symbolized these as having only the characters *A* and *B*. Thus, *AB* gives the character structure of these plants, determined as a result of breeding experiments. For this reason his grouping of symbols is also referred to as the **breeding structure** of the plants.

Sixty-five other plants (line 40), grown from the 315 round yellow seeds and self-fertilized, produced some seeds that were round and yellow and some that were round and green. From these data Mendel concluded that the round yellow seeds from which these plants were grown were true-breeding for the shape trait and hybrid for the color trait. Thus, he symbolized them as *A* for the shape trait and *Bb* for the color trait. This gave these plants the character structure *ABb* and indicated that when self-fertilized they would give two kinds of seeds, round yellow ones and round green ones.

The next 60 plants grown from the 315 round yellow seeds (line 41) also produced two kinds of seeds when self-fertilized, but this time they were round yellow ones and angular yellow ones. This indicates that these plants were hybrids for the shape trait and true-breeding for the color trait, reversing the pattern for the 65 plants just considered. The breeding structure for these plants must have been *Aa* for the shape trait and *B* for the color trait, giving them the character structure *AaB*.

By this time it should be evident that the remaining 138 plants (line 42) were hybrid for both traits; that is, they are dihybrids and should produce four types of seeds: round yellow ones, round green ones, angular yellow ones, and angular green ones. The symbols used to represent them should reflect the hybrid shape trait *Aa* and the hybrid color trait *Bb,* and the total breeding structure for this class must have been *AaBb*. Observed results agreed with these predictions.

Mendel continued with plants grown from 96 (lines 44–46) of the angular yellow seeds. Of these, 28 bore only angular yellow seeds. Since only one shape trait and only one color trait were represented in the seeds, the seeds were true-breeding for each trait. Since the character for the shape trait was recessive, the symbol for it had to be the letter *a*. And since the seeds showed only the dominant character of the color trait, yellow, the correct symbol for this should be a capital *B*. The breeding structure of this type of the 28 plants must therefore have been *aB*. The remaining 68 plants produced both angular yellow and angular green seeds. From this Mendel concluded that these 68 plants must be true-breeding for the shape trait and hybrid for the color trait. Since the character of the shape trait was recessive, the appropriate symbol was the small *a*. Since the characters of the color trait were both the dominant yellow (symbol *B*) and the recessive green (symbol *b*) the breeding structure for these plants must have been *aBb,* as Mendel wrote.

Of the 102 plants (lines 47–49) grown from the round green seeds, he found that 35 produced only round green seeds. Since there was only this one type of seeds produced, the plants must have been true-breeding for each trait. As you can see, the symbol Mendel used to indicate their breeding structure was *Ab*. The remaining 67 plants produced two types of seeds, round and green and angular and green. They must, therefore, have been hybrid for the shape trait and true-breeding for the seed-color trait. Since round is the dominating character for seed shape and angular the recessive, Mendel used the symbol combination *Aa* for this trait. Since all seeds were green, the recessive character for this trait, Mendel used only the symbol *b* for this trait. The breeding structure he gave for these plants, then, was *Aab*.

The 30 plants (line 50) raised from the 108 angular green seeds produced plants having only angular green seeds. To these plants Mendel assigned the symbol combination *ab*. This completed Mendel's analysis of the seeds from the dihybrid cross.

In summary then, the dihybrid seeds from the parental cross were all round and yellow and must have had the breeding structure *AaBb*. When self-fertilized they produced four types of seeds. These in turn gave rise to the various plants of the seond generation from the hybrid whose distribution into nine classes we have just followed, along with Mendel's method of assigning the appropriate symbolic breeding structure to each class.

We should emphasize once agin that Mendel's methods of designing experiments and interpreting his results were completely new in the biology of his time. For this reason it is at least as important to understand his methods of reasoning as it is to know what his final results were. In this section we are able, in a manner of speaking, to look over Mendel's shoulder and watch a great scientist at work: to watch as he transforms the raw data of numbers and kinds of peas, first into empirical laws and second into a quantitative theory of hybridization. This capacity to penetrate behind the data and see the meaning hidden there and bring it forth is the hallmark of a great scientist, just as the hallmark of a great sculptor is the capacity to see the statue in a block of stone and then bring it forth—as Michelangelo did with his *David*.

In lines 55 to 63, Mendel arranges these classes in order, with the four classes that were true-breeding for both traits first (lines 55–58), followed by the four classes true-breeding for one character and hybrid for the other (lines 59–62). The last class in the list is that one hybrid for both traits. Mendel then shows that each form in the first group occurred, on the average, 33 times (line 67); each form in the second class occurred, on the average, 65 times (line 70); and each form in the third class occurred, on the average, 138 times (line 71). These are in the actual proportion of 33:65:138. By dividing each number in the set by the smallest, this proportion reduces to 1.00:1.97:4.18, which is a close approximation of 1:2:4. Had there been no experimental errors, the value for the groups would have been in the proportion of 33:66:132. Then, in lines 77 to 84, Mendel shows that this same 1:2:4 proportion corresponded in the data he had just summarized in the tables (lines 55–63). Each of the first four classes occured relatively only once. Each of the second four classes occurred relatively only twice, while the last class occurred relatively four times among the progeny (on line 85). He then wrote out the complete series of progeny in a way that displays the numerical relationships he had just shown to exist in the data. He also pointed out that this series could be obtained mathematically if the developmentally simple series for the *A* trait was combined, term by term, with the same series for the *B* trait. In this combination series each of the first four combinations, *AB*, *Ab*, *aB*, *ab*,

Table 8.1. Decision Table For Dihybrids

Seeds		Shapes			Colors			Shapes		Colors		Symbols		First Generation From Hybrid
Number	Appearance	Number	Breeding True	Hybrid	Number	Breeding True	Hybrid	Dominant	Recessive	Dominant	Recessive	Shape	Color	
38	All round & yellow	1	x		1	x		x		x		A	B	AB
65	Round yellow & round green	1	x		2		x	x		x	x	A	Bb	ABb
60	Round yellow & angular yellow	2		x	1	x		x	x	x		Aa	B	AaB
138	Round yellow, round green, angular yellow, angular green	2		x	2		x	x	x	x	x	Aa	Bb	AaBb
28	Angular yellow	1	x		1	x			x	x		a	B	aB
68	Angular yellow, angular green	1	x		2		x		x	x	x	a	Bb	aBb
38	Round green	1	x		1	x		x			x	A	b	Ab
67	Round green, angular green	2		x	1	x		x	x		x	Aa	b	Aab
30	Angular green	1	x		1	x			x		x	a	b	ab

occurred only once. Each of the second four combinations, *ABb, aBb, AaB, Aab* occurred twice. The last combination, *AaBb,* occurred four times. We have summarized the entire discussion of Mendel's analysis of the results of the dihybrids in table 8.1.

It is evident that in this expression (line 85) there are no names of traits or characters, no description of appearances of plants or seeds, only symbols and signs. To read the combination series we must understand what the symbols mean and what their groupings in each of the various ways mean. This collection of abstract symbols arranged in certain groupings is very different from the verbal descriptions of plants and their traits and characters with which Mendel started. The whole sequence of events from selecting seeds, planting them, fertilizing the blossoms, watching the pods and seeds develop and ripen, to collecting, classifying, and counting the seeds has culmimated in this dry, compressed series of numbers and letters. Although these results have been derived from data obtained from two particular seed traits, they will state a set of relationships that will hold equally true for any pairing of the seven traits Mendel has been using.

While we are considering Mendel's results and his symbolic notation a word of caution may be in order, especially if you have had some previous contact with what we now call **Mendelian genetics.** If you have, you may have regarded these letters as symbols for **genes,** the units of heredity, and the combinations of symbols as **genotypes.** *This is not correct and must be avoided.* The letter symbols that Mendel used represent the characters present in each member of each class, *not genes.* Characters are observable, genes are not. Genes were conceived after 1900 as invisible particles present in pairs in all cells of a plant or animal except in their sex cells, which contain only one gene of each pair, either the paternal or the maternal. These particles, which carry genetic information from parents to offspring, recombine in pairs in one cell, the zygote, or fertilized egg, which develops into the new individual. In Mendelian genetics each gene is symbolized by one letter. So, for each pair of genes, the hereditary makeup of an organism is written *AA, Aa,* or *aa,* but the makeup of its **gametes** (sex cells) is written *A* or *a.* As you can see, this notation differs decidedly from Mendel's.

We continue now with Mendel's text of section 8 that deals with the trihybrid experiments. This is block B of this section.

Mendel's Text
Block B: The Trihybrid Experiments

Second Experiment:

ABC seed plant,	*abc* pollen plant.
A shape round,	*a* shape angular.
B albumen yellow,	*b* albumen green.
C seed coat grey-brown,	*c* seed coat white.

This experiment was conducted in a manner quite similar to that used in the preceding one. Of all experiments it required the most time and effort. From 24 hybrids a total of 687 seeds was obtained, all of which were spotted, colored grey-brown or grey-green, and round or angular. Of the plants grown from them, 639 bore fruit in the following year and, as further investigations showed, they comprised:

8	plants	*ABC*	22	plants	*ABCc*	45	plants	*AaBbCc*
14	"	*ABc*	17	"	*AbCc*	36	"	*aBbCc*
9	"	*AbC*	25	"	*aBCc*	38	"	*AaBCc*
11	"	*Abc*	20	"	*abCc*	40	"	*AabCc*
8	"	*aBC*	15	"	*ABbC*	49	"	*AaBbC*
10	"	*aBc*	18	"	*ABbc*	48	"	*AaBbc*
10	"	*abC*	19	"	*aBbC*			
7	"	*abc*	24	"	*aBbc*			
			14	"	*AaBC*	78	"	*AaBbCc*
			18	"	*AaBc*			
			20	"	*AabC*			
			16	"	*Aabc*			

The series comprises 27 members. Of these 8 are constant for all traits, and each occurs 10 times on the average; 12 are constant for two traits, hybrid for the third; each appears 19 times on the average; 6 are constant for one trait, hybrid for the other two; each of these turns up 43 times on the average; one form occurs 78 times and is hybrid for all traits. The ratios 10:19:43:78 approach the ratios 10:20:40:80, or 1:2:4:8, so closely that the latter doubtlessly represents the correct values.

The development of hybrids whose parents differ in three traits thus takes place in accord with the expression:

ABC + *ABc* + *AbC* + *Abc* + *aBC* + *aBc* + *aBC* + *abc* +
2*ABCc* + 2*AbCc* + 2*aBCc* + 2*abCc* + 2*ABbC* + 2*ABbc* +
2*aBbC* + 2*aBbc* + 2*AaBC* + 2*AaBc* + 2*AabC* + 2*Aabc* +
4*ABbCc* + 4*aBbCc* + 4*AaBCc* + 4*AabCc* + 4*AaBbC* + 4*AaBbc*
8*AaBbCc*.

Here, too, is a combination series in which the series for traits *A* and *a, B* and *b, C* and *c* are combined with each other. The expressions:

A + 2*Aa* + *a*
B + 2*Bb* + *b*
C + 2*Cc* + *c*

supply all the terms of the series. The constant associations encountered in it correspond to all possible combinations of the traits *A, B, C, a, b, c;* two of them, *ABC* and *abc,* resemble the two parental plants.

In addition, several more experiments were carried out with a smaller number of experimental plants in which the remaining traits were combined by twos or threes in hybrid fashion; all gave approximately equal results. Therefore there can be no doubt that for all traits included in the experiment this statement is valid: *The progeny of hybrids in which several essentially different traits are united represent the terms of a combination series in which the series for each pair of differing traits are combined.* This also shows at the same time that *the behavior of each pair of differing traits in a hybrid association is independent of all other differences in the two parental plants.*

If *n* designates the number of characteristic differences in the two parental plants, then 3^n is the number of terms in the combination series, 4^n the number of individuals that belong to the series, and 2^n the number of combinations that remain constant. For instance, when the parental types differ in four traits the series contains $3^4 = 81$ terms, $4^4 = 256$ individuals, and $2^4 = 16$ constant forms; stated differently, among each 256 offspring of hybrids there are 81 different combinations, 16 of which are constant.

All constant associations possible in *Pisum* through combination of the above-mentioned seven characteristic traits were actually obtained through repeated crossing. Their number is given by $2^7 = 128$. At the same time this furnishes factual proof that *constant traits occurring in differing forms of a plant kindred can, by means of repeated artificial*

160 *fertilization, enter into all the associations possible within the rules of combination.*

Experiments on the flowering time of hybrids are not yet finished. However, it can already be reported that it is almost exactly intermediate between that of the seed plant and that of the pollen plant,
165 and that development of the hybrids probably proceeds in the same manner with respect to this trait as it does for the remaining traits. The forms chosen for experiments of this nature must differ by at least 20 days in the mean date of blooming; it is also necessary that all the seeds be planted at an equal depth in order to achieve simultaneous germination,
170 and, furthermore, that throughout the flowering period any relatively large temperature fluctuations with consequent acceleration or delay in blooming be taken into account. It is obvious that this experiment has various difficulties to overcome and demands great attention.

When we try to summarize briefly the results obtained, we find
175 that those differing traits that permit easy and certain differentiation in the experimental plants *show completely concordant behavior in hybrid association.* Half of the progeny of plants hybrid in one pair of differing traits are hybrid again, while the other half become constant, with the characteristics of the seed and pollen plants in equal
180 proportion. When, through fertilization, several differing traits become united in a hybrid, its progeny represent the terms of a combination series, in which the series of any one pair of differing traits are combined.

The complete agreement shown by all characteristics tested
185 probably permits and justifies the assumption that the same behavior can be attributed also to the traits which show less distinctly in the plants, and could therefore not be included in the individual experiments. An experiment on flower stems of different lengths gave on the whole a rather satisfactory result, although distinction and
190 classification of the forms could not be accomplished with the certainty that is indispensable for correct experiments.

Interpretive Comments

In his trihybrid experiments Mendel added a third character, seed coat color, to the other seed characters that he used in his dihybrid experiments. At first, such a choice for a third trait seemed to be a logical one, especially since he was already dealing with two other seed traits.

But this choice involved two main difficulties, which rendered Mendel's work harder than he might have expected.

The first difficulty is the identification of the color of the cotyledons. In this experiment, the seed coats of all the seeds he was dealing with were an opaque grey-brown color. As a result, he could not see the colors of the cotyledons. In order to determine their color, he had to devise some method of opening the seed coat that did not damage the embryo inside. However, here again, he did not tell us how he did it. As we shall see, he adopted an indirect method.

The second difficulty is that the seed coat, as we mentioned before, is not part of the seed itself, but part of the plant that bears the seed. This presents a problem for the plant breeder because the coat color of a seed cannot be directly associated with the other two seed traits. To determine the seed coat color, which is associated with the other two seed traits, Mendel had to plant the hybrid seed, grow the plant from that seed to maturity, and then determine the color of the seed coat of the seeds that this plant bore. But again Mendel never tells us any of this. He tells us only that the hybrid seeds produced 639 plants, which formed seeds the following year (line 101), and as "further investigations showed, they comprised" the data in the table (lines 102–114) That statement, "further investigations showed," hides a multitude of difficulties, which we cannot discuss here in the text. If you want to see what the magnitude of his problems in record keeping and interpretation were, we have worked them out and placed them in appendix D.

Let us note here that Mendel could have made this experiment much less difficult to carry out had he used three traits, all of which matured either at the same time or at least in the same growing season. Any three of the last four traits—shape of the ripe pod, color of the unripe pod, position of flowers, difference in stem length—would have worked much better. Although he did not tell us why he chose the three traits he did, we might speculate that being an experienced plant breeder he knew what the problems would be, knew he could solve them, and so went ahead. It may be that he enjoyed the challenge and welcomed a chance to demonstrate his abilities. Or he may have felt this difficult demonstration would be exceptionally convincing evidence of the correctness of his conceptions. At any rate, it was a virtuoso display of his skill and ability as a master plant breeder exulting in the full exercise of his powers. It was also a happy display of his power as a scientist whose intuition had led him to find a new approach to the study of plant hybrids, blending biology, the physical sciences, and mathematics. This blending enabled him eventually to find a new natural law that gave both qualitative and quantitative explanations for many long-known phenomena in this area and enabled him to predict correctly the results of new experiments.

Now, leaving for the strong of heart further discussion of the difficulties Mendel had in carrying this trihybrid experiment to completion, let us turn again to Mendel's text.

Mendel began the trihybrid part of this section by giving the character structures of the seed plant and the pollen plant (lines 93–96), but did not give us any information about the hybrid seeds produced. However, from what we already know we can infer that these seeds were round and had yellow cotyledons and grey-brown seed coats. They were round because round is dominating over angular seed shape; the cotyledons were yellow because yellow is dominating over green cotyledon color; and the seed coat was grey-brown. As you remenber, this is because the seed coat is part of the seed plant, which in this case, Mendel tells us, is true-breeding for grey-brown seed coat.

Since the experiment was carried out in much the same fashion as the dihybrid experiment (lines 97–98), we know he must have planted these seeds and raised the next generation, the first generation from the hybrids. He did not tell us how many of the hybrid seeds he planted. However, he did tell us (lines 99–100) that he collected 687 seeds from 24 hybrid plants. According to Mendel, all of these seeds had the same seed coat color (grey-brown or grey-green) and they were either round or angular. From what follows we know that Mendel recorded the color of the cotyledons of those seeds, but he did not tell us what they were or how he determined them. However, from what we know of the results of the dihybrid experiments, we can at least be sure that they were either yellow or green.

Let us turn our attention to the data Mendel accumulated in the table in lines 103–114. These data correspond to the dihybrid data Mendel reported in lines 55–63 and were obtained, as Mendel said, in essentially the same way. The first column at the left of the trihybrid data consists of all the true-breeding members. These eight members correspond to the first four in the dihybrid data. Two of them, *ABC* and *abc*, have the same forms as the parents of the hybrids, while the remaining six consist of new true-breeding forms produced by recombinations of the six parental characters. There are 77 plants distributed over 8 forms for an average of 9.625 plants per form or, rounding off, 10 per form.

The second column of the trihybrid data consists of forms constant for two traits and variable for one. Though the trihybrid data includes more forms and forms of greater complexity than the dihybrid data, they parallel the second set of forms of progeny in the dihybrid data. In this column there are 228 plants distributed over 12 forms, giving an average of 19 plants per form.

In column three, the first six forms are constant for one trait and variable for two. There are no data for the dihybrid experiment that parallel these. In this column there are 256 plants distributed over the first six groups, giving an average of 42.7 plants per group. Or, rounding, we have 43 plants per group.

The remaining entry in this column is variable in all three traits. This corresponds within the limits of numbers to the last entry in the dihybrid data, which consisted of plants variable in both traits. In the trihybrid data there are 78 plants in this last group.

Mendel summarized the ratios for the dihybrids as 1:2:4. In the trihybrids the summary (line 120) is 1:2:4:8, which he obtained, as he did the 1:2:4 for dihybrids, by dividing each term in the experimentally determined series by the smallest value in the series. This gave 1 to 1.9 to 4.3. to 7.8. When these were rounded to the nearest whole number, Mendel obtained the 1:2:4:8 series of line 120.

Again in parallel to his treatment of the dihybrids, these relative numbers provided the coefficients for the forms in the columns and enabled Mendel to write out the combination series formula for the progeny of this cross where the parents differed in three traits (lines 124–128). As in the case of the dihybrid cross, Mendel recognized that this long, experimentally determined series could be generated theoretically or mathematically through the combination, term-by-term, of the appropriate single-trait series.

It is entirely possible that Mendel's main interest in this cross was discovering whether the experimental results would match, term-by-term, the series he could generate by combining the three simple developmental series for the pairs of characters used (lines 131–134). That they did match indicated that this was a way of predicting the series for crosses involving any number of pairs so long as they were of the types he used in his experiments. In lines 137–140 he writes that he verified the correctness of his idea with other combinations, but again does not tell us which ones.

Mendel summed up the results of his experiments with two- or three-trait hybrids in the statement in lines 141–144, which is his combination series law. Then, almost as an afterthought, in lines 144–146, he answered affirmatively the question posed at the beginning of this section (lines 2–5): "The next task consisted in investigating whether the law of development thus found would also apply to a pair of differing traits when several different characteristics are united in the hybrid through fertilization." This statement of the independence of traits is another empirical law based on concordant data from experiments.

Having settled the main matters of concern to him he then turned to one of those little mathematical disgressions he seems to have enjoyed. We have summarized his demonstration in lines 147–154 in tabular form:

Number of traits being followed	Number of constant forms in progeny	Number of terms in combination series	Number of individuals in series
1	$2 = 2^1$	$3 = 3^1$	$4 = 4^1$
2	$4 = 2^2$	$9 = 3^2$	$16 = 4^2$
3	$8 = 2^3$	$27 = 3^3$	$64 = 4^3$
n	2^n	3^n	4^n

Having finished his mathematical diversion Mendel made one of his brief statements (lines 155–161) which, like a partly opened door, give us a glimpse of what must have been another large series of experiments. All we know is that he writes that he produced all the constant associations of his chosen traits by making crosses and that there were 128 of them. Although Mendel does not comment on this here, these experiments gave a further example of the law of independence of traits stated in lines 144–146.

He then reports still another set of experiments involving a different trait of his peas, flowering time. He apparently had two varieties, which took quite different amounts of time to produce their flowers. He reviews the many problems involved in trying to use flowering time as an experimental trait (lines 168–173). When he made the cross of the two parents he found that, while all the hybrid plants were alike in flowering time, that time was not the same as the flowering time of either parent but was "almost exactly intermediate between" those of the parents (lines 163–164). Here was something new: a trait that behaved differently from any of the seven traits previously used in his hybridizing experiments. In all of those, one of the contrasting characters of the traits was dominating, the other recessive. In the flowering-time experiment neither character was dominating or recessive. What, then, would have happened had Mendel continued the experiment and allowed the hybrids to self-fertilize? Mendel offers the supposition that the development would have proceeded just as in the other seven traits. This suggests that he had not finished this experiment before writing his report. We now know that if Mendel had carried out this experiment he would *not* have obtained his familiar 3:1 (3 dominant to 1 recessive) ratio in the first generation from the hybrids because neither character was dominant. Instead, he would have found that the progeny were produced in a 1:2:1 ratio (1 true-breeding early flowering like one of the parents, 2 intermediate flowering like the hybrid, 1 true-breeding late flowering like the other parent). Because in this case there is no dominance and there are three distinct characters instead of two, the breeding ratio 1:2:1, so critical in Mendel's reasoning, would not only have been found, but would have been found one generation sooner than in his other experiments. The reason for this is that the lack of dominance in flowering time permits a clear distinction between the parental characters and the hybrid character. Some of the minor empirical laws he had found earlier would have been missing in this case, but the master law, the combination series law, would have emerged in the same form as it did.[1] This brief note about Mendel's results with the flowering-time experiments shows clearly that he was aware that the dominating-recessive relationship was not universal.[2]

Notes

1. It is commonly believed that the 3:1 ratio is the main discovery of Mendel and the basic idea of Mendelism, but as the reader is now aware it was the 1:2:1 ratio that led Mendel to his understanding of the development of hybrids and the laws that govern their development. It was also the 1:2:1 ratio that led him to formulate a one-pollen-grain-to-one-egg-cell mechanism of fertilization, which explained the empirical laws he had discovered. Later, researchers showed that this mechanism was correct.

2. Mendel's comments are interesting for the following reason. The law of dominance, which states that the hybrid always has the characteristic of one of the parents, has been attributed to Mendel, possibly because all of the seven

traits for which he gives data show dominance. But as we see here, he was perfectly aware that other traits do not show dominance (that hybrids could be intermediate in their appearance). The man responsible for stating such a law was Hugo de Vries, one of the three men who were responsible for resurrecting Mendel's paper in 1900.

§9

The Reproductive Cells of Hybrids

Thus far Mendel has been concerned with finding the "generally applicable laws of the formation and development of hybrids." In the course of this search he found a series of empirical natural laws, each of which describes quantitatively some part of the behavior of his hybrids. He discovered these laws by using a pattern of thinking known as inductive reasoning. In forming his monohybrids he found that the progeny of any one cross were all alike in showing the character of the dominating parent. This shared pattern of behavior became the basis for an empirical law. When the hybrids of each trait reproduced by self-fertilization, each produced two classes of offspring very nearly in the constant ratio of 3 of the dominating form to 1 of the recessive form. Here was a second pattern of behavior running across all of the experimental traits. This pattern became the basis for another empirical law. He then repeated this process with the second generation from the hybrids and found another pattern shared in common in the progeny (1 of the dominating true-breeding form, 2 of the dominating hybrid form, to 1 of the recessive true-breeding form) and thus found another empirical natural law. In each generation the empirical law bound together a series of separate individual facts into a relationship that represented them all. Not only did it represent all of the examples on which it was based, but it represented equally well the behaviors of hybrids of any trait that conformed to the criteria Mendel had chosen in selecting those used in his experiments. The empirical laws are thus powerful tools for *predicting* what would happen in experiments that had not yet been tried.

In the section on Mendel's experiments with dihybrids and trihybrids, which you have just read, Mendel employed this same method of reasoning. When he completed the analysis of his data from the dihybrid experiments he found a more general law than he had with his monohybrids, his combination series law. This master law bound together in a single relationship all the individual subordinate empirical laws he had found. The method of term-by-term combination of the simple developmental series for each trait, which gave the developmental series for the dihybrids, worked equally well for the trihybrid cross. This method now enabled Mendel to calculate what the series should be for the progeny of other dihybrids, trihybrids, or polyhybrids so long as the individual traits conformed to his criteria.

At this point Mendel found that "generally applicable law of the formation of development of hybrids" for which he was searching. We have seen how the repeated application of the pattern of inductive reasoning was crucial in obtaining his results. Before we move to section 9 there are a few things we need to keep in mind about what we have been studying.

First, the laws of hybrids Mendel had discovered were *empirical* laws. They were derived from the data produced by his experiments, not from a theory of the formation of hybrids that Mendel already had in his mind. The experiments were designed to discover the basic facts and relationships from which the laws could be found. Note that if, on the other hand, the laws had been derived from a theory, they would be referred to as *theoretical* laws.

Second, the laws that Mendel found were *quantitative* laws. They specified not only what types of offspring would be produced, but also the relative amounts in which the various types would be produced. Laws of this type were common in the physical sciences of Mendel's time but very rare in the biological sciences.

Third, the laws were concerned with traits and characters represented by abstract letter symbols. The laws themselves were written partially in words and partially in symbols. Mendel developed this system of notation to make thinking and writing about hybrids and hybridization simpler, easier, and more reliable once the notation had been mastered.

Having discovered the empirical laws of hybrids, Mendel's next task was to propose and test a theoretical scheme that would, first of all, explain in as simple a fashion as possible how hybridization took place and why the laws he had discovered took the form they did. Second, the laws would have to appear reasonable and useful to other hybridizers. To appear reasonable to other hybridizers, the scheme should employ methods and material familiar to them. It would have to suggest experiments that could be used to test its correctness. It should enable the experimenter to make reliable predictions about what forms of offspring should be produced in the experiments and the relative amounts in which they should be produced.

The creation and testing of such a scheme, or theory, required a different pattern of thinking from the one Mendel had been using in searching for his laws.[1] In fact, the pattern he uses in this section is almost the exact reverse of the earlier one. In this section he makes a small number of assumptions about what happens when plants reproduce sexually. And these assumptions should apply, not just to the plants he was using in his experiments but to sexually reproducing plants in general.[2] These assumptions went well beyond the simple observable facts of pollen and egg formation and their combination in fertilization and were concerned with things and processes hidden from the eye.

Next he had to ask himself how he might test whether these assumptions were true or false. This involved designing experiments in his mind and thinking about what the results should be if the ideas were true or if they were false. The pattern involved has the following form: if the idea is true, then the results of this particular breeding cross ought to give me a test of its correctness. On the basis of my idea I can predict what those results should be. If the results of my experiments do not agree with my predictions, then I will have to modify some of the assumptions or give the idea up and invent another if I can. This is the classic deductive pattern of reasoning used both in building connections from known laws to a theoretical explanation and in testing the value of the explanation.

Having come to a conclusion about what sort of cross might test one or more assumptions and having made some predictions, Mendel took the next step: he designed one or more special experiments and actually carried them through to see what data would be obtained. Mendel did this twice in section 9 and then tested his ideas on the data from his monohybrid and dihybrid crosses.

In each of these he predicted in detail what the results of the experiment should be, both qualitatively and quantitatively. In each case the results of experimentation agreed very closely with his predictions. Since they did so, he felt that he had proved the correctness of his hypotheses about the formation and composition of the germinal and pollen cells of hybrids and also the pattern of their union in fertilization. Having tested his proposed process on these specially designed experiments, he then applied it to his monohybrid, dihybrid, and trihybrid experiments. Here again the theoretical process or mechanism led to results in full agreement with the results of the earlier experiments.

Just a word more here before you begin section 9. This is one of the most difficult sections in the whole of Mendel's paper. It is doubtful that many of Mendel's contemporaries worked their way through it and understood it. This is probably true also for two of those researchers who reintroduced Mendel's work to biologists and hybridizers in 1900. Therefore, if you find that this section is hard to understand, you have plenty of company. So good luck and have at it. In this section we have again divided Mendel's text and inserted interpretive comments as seemed appropriate.

Mendel's Text
Block A: Assumptions and Deductions

1 The results of the previously cited investigations suggested further experiments whose outcome would throw light on the composition of seed and pollen cells in hybrids. An important clue is the fact that in *Pisum* constant forms appear among the progeny of hybrids and that they do so in
5 all combinations of the associated traits. In our experience we find everywhere confirmation that constant progeny can be formed only when germinal cells and fertilizing pollen are alike, both endowed with the potential for creating identical individuals, as in normal fertilization of pure strains. Therefore we must consider it inevitable that in a hybrid
10 plant also identical factors are acting together in the production of constant forms. Since the different constant forms are produced in a *single* plant, even in just a *single* flower, it seems logical to conclude that in the ovaries of hybrids as many kinds of germinal cells (germinal vesicles), and in the anthers as many kinds of pollen cells are formed as
15 there are possibities for *constant* combination forms and that these germinal and pollen cells correspond in their internal make-up to the individual forms.

Indeed, it can be shown theoretically that this assumption would be entirely adequate to explain the development of hybrids in separate
20 generations if one could assume at the same time that the different kinds of germinal and pollen cells of a hybrid are produced on the average in equal numbers.

In order to test this hypothesis experimentally, the following

experiments were chosen. Two forms which differed constantly in seed shape
25 and albumen coloration were combined by fertilization.

If the differing traits are again designated by *A, B, a, b,* then:

AB seed plant,	*ab* pollen plant,
A shape round,	*a* shape angular,
B albumen yellow,	*b* albumen green.

30 The artificially fertilized seeds were sown, together with several seeds of the two parental plants, and the most vigorous specimens were chosen for reciprocal crosses. Fertilized were:

1. The hybrid with pollen from *AB.*
2. The hybrid " " " *ab.*
35 3. *AB* " " " the hybrid.
4. *ab* " " " the hybrid.

From each of these four experiments all flowers on three plants were fertilized. If the above assumption were correct, then germinal and pollen cells of the forms *AB, Ab, aB, ab* should develop in the hybrids,
40 and combined would be:

1. *Germinal cells AB, Ab, aB, ab* with pollen cells *AB.*
2. " " *AB, Ab, aB, ab* " " " *ab*
3. " " *AB* " " " *AB, Ab, aB, ab.*
4. " " *ab* " " " *AB, Ab, aB, ab.*

45 From each of these experiments, therefore, only the following forms could result:

1. *AB, ABb, AaB, AaBb.*
2. *AaBb, Aab, aBb, ab.*
3. *AB, ABb, AaB, AaBb.*
50 4. *AaBb, Aab, aBb, ab.*

Furthermore, if the individual forms of the hybrid's germinal and pollen cells are, on the average, formed in equal numbers, then in each experiment the four combinations listed must stand in equal ratio to each other. Complete agreement of the numerical values was not to be expected,
55 however, since in every fertilization, even in a normal one, some germinal

cells fail to develop or die later, and even some well-developed seeds do
not succeed in germinating after being planted. Moreover, the assumption
requires only that there be a tendency to approach equality in the number
of different kinds of germinal and pollen cells produced, yet this number
60 does not have to be attained by each hybrid with mathematical exactness.

The primary objective of the *first* and *second* experiments was to
test the composition of hybrid germinal cells; that of the *third* and
fourth experiments was to determine the composition of pollen cells. As
shown by the above compilation, the first and third experiments, like the
65 second and fourth, had to give identical combinations. The results should
have been partly observable in the second year in the shape and
coloration of the artificially fertilized seeds. In the first and third
experiments the dominating traits of shape and color, *A* and *B,* occur in
every combination, constant in some, in hybrid association with the
70 recessive characters *a* and *b* in others, and must, for that reason, imprint
their characteristics on all seeds. Therefore, if the premise were
correct, all seeds should appear round and yellow. In the second and
fourth experiments, on the other hand, one association is hybrid in shape
and color, and therefore the seeds are round and yellow; another is
75 hybrid in shape and constant in the recessive trait of color, therefore
the seeds are round and green. The third association is constant in the
recessive trait of shape and hybrid in color, therefore the seeds are
angular and yellow. The fourth is constant in both recessive traits;
therefore the seeds are angular and green. Hence, one could expect four
80 different kinds of seeds from these two experiments, namely: round
yellow, round green, angular yellow, angular green. The yield was in
complete agreement with these expectations. There were obtained in the

First experiment, 98 exclusively round yellow seeds;
Third " 94 " " " " ;
85 Second experiment, 31 round yellow, 26 round green, 27 angular yellow,
26 angular green seeds;
Fourth experiment, 24 round yellow, 25 round green, 22 angular yellow,
27 angular green seeds.

Interpretive Comments

The clue that enabled Mendel to begin constructing his explanatory scheme was the production of true-breeding, or constant, progeny when his hybrids reproduced by self-fertilization. In the monohybrid experiments he found that he could recover both true-breeding parental forms. In the dihybrid and trihybrid experiments he recovered not only the true-breeding parental forms but also unexpectedly some new combinations of the characters, which did not exist in the parental plants but which were also true-breeding. He knew that the production of these various true-breeding forms had to be the result of the union of egg and pollen cells in fertilization. But so did the production of the various hybrid progeny. The problem was to find out, if possible, what kinds of egg and pollen cells were produced in each cross and in what numbers.

Mendel also knew that if the parental plants of his monohybrids were allowed to self-fertilize, all the progeny of each type would be exactly like the parent plant from which they came. This was, of course, what was meant by saying that the plants were true-breeding. Since, in such a fertilization, both the egg and pollen cells were produced in the same plant, they must be alike in regard to the characters present in them (lines 5–9). Mendel reasoned that if, in this case, the identical true-breeding plants were produced by self-fertilizing the hybrids and also by self fertilizing the parent plants, the hybrids must produce germinal and pollen cells like those produced by the individual true-breeding parents. His argument took the following form: if the *same* true-breeding form is produced in both cases, then the same process must be taking place in both—that is, the union of identical pollen and egg cells.

In the self-fertilization of the monohybrids, type *A* germinal and pollen cells must be produced and then united in fertilization to reproduce the true-breeding type *A* parent among the progeny of the cross. In the same way, type *a* germinal and pollen cells must be produced in the monohybrid and then united in fertilization to reproduce the true-breeding type *a* parent among the progeny of the cross. The final clue was stated by Mendel in lines 11 and 12, that the different constant forms were produced, not just in the seeds from a single plant with many flowers but in the seeds from a single flower. He had narrowed the area within which the phenomena of hybrid production took place to the individual flower with its ovary and anthers.

These clues enabled Mendel to propose a series of assumptions, which provided his framework for explaining the process of formation of the progeny of hybrids. That basic framework consisted of four assumptions that he stated explicitly and a fifth one he did not. They were as follows (see lines 13–22 in his text):

Assumption 1. In the *ovaries* of hybrids as many kinds of germinal cells (line 13) are produced as there are possibilities for constant combinations of traits in their progeny (line 13).

Assumption 2. In the *anthers* of hybrids as many kinds of pollen cells are produced as there are possibilities for constant combinations of traits in their progeny (lines 14–15).

Assumption 3. The combinations of traits present in the germinal and pollen cells correspond to the combinations of traits present in the constant forms in the progeny (lines 16–17).

Assumption 4. In a hybrid, the different kinds of germinal and pollen cells are produced, *on the average,* in equal numbers (lines 20–22).

Assumption 5. Fertilization occurs equally often in each possible combination.

In lines 23–29 Mendel describes how he produced the special hybrid he intended to use in testing his proposed scheme. He used a true-breeding plant producing round yellow seeds (*AB*) as the seed plant and a true-breeding plant producing angular green seeds (*ab*) as the pollen plant. He probably fertilized only a few of the flowers of the seed plant with pollen from the other parent plant. The rest of the flowers were allowed to self-fertilize. Thus, the parental seed plant produced two kinds of seeds: hybrid seed from the artificially fertilized flowers and non-hybrid seed from the self-fertilized flowers. These were harvested in the fall and the seed from each flower kept separate. The pollen parent also self-fertilized and yielded its regular seeds.

The next fall some of the hybrid seeds were planted in the garden and also some of the self-fertilized seeds from each of the parent plants (lines 30–32). When they reached the flowering stage, Mendel made the crosses shown in lines 33–36. These four crosses are called **back-crosses** because the hybrid progeny are crossed back to the parent that produced them.

In the first two the hybrid served as seed parent, thus supplying germinal cells to be fertilized in experiment 1 with pollen from the parent plant, *AB,* and in experiment 2 with pollen from the other parent, *ab.*

In the third and fourth crosses the hybrid served as pollen parent, providing pollen to fertilize the germinal cells of the two parent plants. As you can see, experiment 3 is just the reverse of experiment 1 and experiment 4 is the reverse of experiment 2. Such crosses in which the roles of the two members are reversed are commonly called reciprocal crosses. Mendel's purpose in crosses 1 and 2 (lines 61–62) was to test the composition of the hybrid germinal cells by fertilizing them with two kinds of pollen cells, each of whose character structure (composition) was known. In crosses 3 and 4 his purpose (lines 62–63) was to test the composition of the pollen cells of the hybrid by using them to fertilize two kinds of germinal cells whose internal composition was known. The results of these fertilizations should allow him to test his assumption that the composition of the two kinds of reproductive cells, germinal and pollen, produced by the hybrid was the same.

In any test like this it is essential for the experimenter to predict in as much detail as possible what results are expected. Then, when the actual experiment is performed, the results obtained can be compared with the predictions. In this way the assumption that was the source of the prediction can be tested. Mendel predicted in lines 41 and 42 what the composition of the germinal cells of the hybrid should be and what the composition of the pollen cells of each parent should be. In lines 43 and 44 he predicted what the composition of the germinal cells of the parents and the pollen cells of the hybrid should be. In lines 47 through 50 Mendel set down his predictions for the composition of each of the kinds of offspring that should be produced by each of the crosses in lines 41 through 44. In lines 51 through 54 he predicted the relative numbers in which these offspring should appear in each cross: 1 to 1 to 1 to 1. Mendel commented in lines 54 through 60 on some of the factors that might cause problems in single experiments and offered suggestions as to how these problems could be solved.

Starting in line 65 Mendel gave specific predictions of what should happen in each experiment, with a brief explanation of why. We have given these in table 9.1.

Thus far, Mendel's ideas about how the progeny of hybrids should look appear to be right on target. The predicted appearances were observed and occurred in the relative numbers predicted from his scheme or theory. However, this confirmation was not enough to satisfy Mendel. In order to really establish the correctness of his ideas, he needed to demonstrate that the 98 round yellow seeds obtained in experiment 1 and the 94 round yellow seeds obtained in experiment 3 actually had

Table 9.1. Abbreviations for Appearance of Progeny

Predicted Character Structures (lines 47–50)	Predicted Appearance (lines 65–81)	Observed Appearance (lines 82–88)
1.* AB, ABb, AaB, AaBb	all RY	98 RY
2. AaBb, AaB, aBb, ab	1 RY : 1 RG : 1 AY : 1 AG	31 RY, 26 RG, 27 AY, 26 AG = 110
3. AB, ABb, AaB, AaBb	all RY	94 RY
4. AaBb, AaB, aBb, ab	1 RY : 1 RG : 1 AY : 1 AG	24 RY, 25 RG, 22 AY, 27 AG = 98

NOTE: RY = round yellow, RG = round green, AY = angular yellow, AG = angular green.
*These numbers correspond to the numbers of the original experiments.

the predicted character structures and that these were produced in the proper proportions. This required producing the next generation by self-fertilization and determining the character structures of each of the seeds. These further tests of experiments 1 and 3 and then 2 and 4 make up the next block of section 9.

Mendel's Text
Block B: Analysis of Results of the Four Experiments

A favorable result could hardly be doubted any longer, but the
90 next generation would have to provide the final decision. From the seeds
sown, in the first experiment 90 plants, and in the third 87 plants, bore
fruits in the following year; these yielded in

	Experiments			
	1.	*3.*		
95	20	25	round yellow seeds	*AB.*
	23	19	round yellow and green seeds	*ABb.*
	25	22	round and angular yellow seeds	*AaB.*
	22	21	round and angular, yellow and green seeds . .	*AaBb.*

In the second and fourth experiments the round yellow seeds produced
100 plants with round and angular, yellow and green seeds . . . *AaBb.*

From the round green seeds plants were obtained with round and angular
green seeds . . . . *Aab.*

The angular yellow seeds produced plants with angular yellow and green
seeds . . . . *aBb.*

105 From the angular green seeds plants were raised which again yielded only angular green seeds . . . . *ab.*

Though in these two experiments also some seeds did not germinate, no change could be effected in the figures found in the preceding year, since each kind of seed produced plants that were, with respect to their
110 seeds, alike among themselves and different from others. Thus in the

Second experiment	*Fourth experiment*	
31	24	plants yielded seeds of form *AaBb.*
26	25	" " " " *Aab.*
115 27	22	" " " " *aBb.*
26	27	" " " " *ab.*

In all experiments, therefore, all forms postulated by the preceding hypothesis appeared, and did so in nearly equal numbers.

Interpretive Comments

Having completed the four experiments and obtained evidence that was favorable, Mendel then completed the testing of his theory by carrying the progeny of each of the four experiments through another generation of selfing. To do this he followed the development of the seeds from the previous year's experiments 1 and 3 separately from those from experiments 2 and 4.

Since the seeds obtained from crosses 1 and 3 were all yellow and round in appearance as predicted, it was necessary to plant them and to produce the next generation by selfing in order to determine whether or not they actually had the character structures predicted in lines 47 and 49 and that these had been produced in approximately equal numbers. When the yellow round seeds had been planted and new seeds were produced, Mendel reported the results in lines 95 through 98. The results were nearly the same for the new experiments 1 and 3. The plants that yielded only round yellow seeds were true-breeding for both shape and color and therefore had the character structure *AB*. Those that produced round yellow and round green seeds were constant for shape and hybrid for color and therefore had the character structure *ABb*. The character structures of each of the remaining groups of plants could be deduced in the same way.

Since the plants from experiments 1 and 3 all had one parent with character structure *AB*, it was possible for Mendel to deduce the composition of the germinal and pollen cells corresponding to each class of seed. The arrangement of the data below shows how this could be determined:

Seed characters observed	Pollen cell character used	Hybrid germinal cell character deduced
AB	AB	AB
ABb	AB	Ab
AaB	AB	aB
AaBb	AB	ab

Since the data for experiment 3 of the second year were essentially the same as those for experiment 1 of that year, Mendel concluded that both the germinal and pollen cells of the original hybrid were produced in the same four classes and in very nearly equal numbers. Thus, experiments 1 and 3 confirmed Mendel's assumptions about the number and character of the germinal and pollen cells produced.

We have summarized the data he obtained from experiments 2 and 4 (lines 85 through 88) in table 9.2. The numbers are those of the seeds. Ignore column 4 for now; we will return to it later.

If we go back to experiment 2, in which the germ cells of the round yellow hybrid were fertilized with pollen from the angular green parent, we can see in line 1 of table 9.2 that the cross produced 31 round yellow seeds. The composition of the germ cells from the RY (round yellow) plant was not known. However, the composition of the pollen cells from the AG (angular green) parent was known. Since both A and G are the recessive characters of the traits, the composition of the hybrid cell in this case can only be *ab*. Therefore, if the cross of the germinal cells of the RY parent with the *ab* pollen cells produced only RY seeds (31 of them), the breeding structure of the germinal cells that produced those seeds must have been *AB*. These *AB* cells were fertilized by the *ab* pollen cells to give seeds of breeding structure *AaBb,* which would be round and yellow since A and B are the dominating traits.

In line 2 of table 9.2, the cross between the RY hybrid and the AG parent produced 26 round green seeds. Again we do not know the breeding structure of the germ cells from the RY hybrid, but we do know the breeding structure of the pollen cells from the AG parent and

Table 9.2. Analysis of Breeding Data for Experiments 2 and 4

1		2	3	4
2d Exp. # of seeds	4th Exp. # of seeds	Appearance of seeds	Appearance of the progeny of these seeds formed via self-fertilization	Breeding structure of the original seeds as deduced by Mendel*
31	24	all round yellow	RY, RG, AY, AG	AaBb
26	25	all round green	RG, AG	Aab
27	22	all angular yellow	AY, AG	ABb
26	27	all angular green	AG	ab

NOTE: RY = round yellow, RG = round green, AY = angular yellow, AG = angular green.

*These breeding structures were deduced by Mendel from the type of cross involved in the production of the seeds in column 3. But they can be deduced more simply from the appearance of the seeds in column 2. See text.

the appearance of the seeds. If the seeds are green they must have received the *b* character from both parents. For the seeds to have been round they must have received the *A* character from the hybrid and an *a* character from the pollen parent. Therefore, the germ cells of the hybrid that produced the round green seeds must have been *Ab*. By following a similar line of reasoning we can deduce the breeding structure of the germ cells of the hybrid that produced the angular yellow and the angular green seeds of experiments 2 and 4.

We have shown in the two paragraphs above how the breeding structure of the germ cells of the hybrid can be deduced from the appearance of the seeds of the hybrid when crossed to its recessive parent. The design of experiments 2 and 4 by Mendel was such that he successfully, and in one stroke, achieved his goal of testing the correctness of his theory of the formation of germinal and pollen cells and the assumption that they were produced on the average in equal numbers (table 9.2). By designing such experiments Mendel invented a type of cross, since called a **testcross,** which is used regularly by plant and animal breeders to determine whether a plant or animal is a hybrid or not. The plant or animal tested is crossed to a plant or animal that shows the recessive trait. If the offspring of a testcross are all the same, the plant or the animal is not a hybrid. If the offspring are not all the same, the animal or plant is a hybrid.

Yet, for unknown reasons, although he had the results he expected, Mendel grew the seeds that the hybrid plant bore through a cycle of self-fertilization (lines 89–90), got the breeding results indicated in column 3, table 9.2, and then deduced the breeding structure of these seeds (column 4, table 9.2) in a manner similar to, but more complex, than the one we described in the two above paragraphs. However, Mendel did not tell us how he did this. We should note that in lines 113–116 he wrote what the breeding structures of the seeds yielded by the plants were. What he meant was that these were the breeding structures of the plants that yielded those seeds. Nevertheless, experiments 1 and 3 and experiments 2 and 4 demonstrate that the hybrid produces identical germinal and pollen cells in equal numbers and that these cells correspond in their internal makeup to the individual constant forms. Mendel could state (lines 117–118) that all his predictions had been verified. However, all these last experiments had been done with only two seed traits. Would essentially the same results be obtained with nonseed traits? To test whether they would, he devised experiments on flower color and stem length.

Mendel's Text
Block C: Second Test of His Predictions

In a further test the traits of *flower color* and *stem length* were
120 included in the experiments and selection was made in such a way that in
the third year of experimentation every trait had to appear in *half* of
all plants if the above assumption were correct. *A, B, a, b* serve again
as designation of the different traits.

A flowers purplish-red, *a* flowers white.
125 *B* stem long, *b* stem short.

Form *Ab* was fertilized by *ab,* producing hybrid *Aab.* In addition, *aB* was
also fertilized with *ab,* yielding hybrid *aBb.* For further fertilization
in the second year the hybrid *Aab* was used as seed plant, the hybrid *aBb*
as pollen plant.

130 Seed plant *Aab*, Pollen plant *aBb*.
Possible germinal cells *Ab*, *ab*, Pollen cells *aB*, *ab*.

From fertilization involving the possible germinal and pollen cells, four combinations had to result, namely:

$$AaBb + aBb + Aab + ab$$

135 From this it becomes apparent that, according to the above assumption, of all plants in the third year of experimentation

Half should have	violet-red flowers (*Aa*)	. . . .	terms	1,3
" " "	white flowers (*a*)	. . . . . .	"	2, 4
" " "	a long stem (*Bb*)	. . . . . .	"	1, 2
140 " " "	a short stem (*b*)	. . . . . .	"	3, 4

Out of 45 fertilizations of the second year, 187 seeds were obtained, from which 166 plants reached the flowering stage in the third year. Among them the individual terms appeared in the following numbers:

Term	*Color of flower*	*Stem*	
145 1	violet-red	long	. . . . 47 times
2	white	long	. . . . 40 "
3	violet-red	short	. . . . 38 "
4	white	short	. . . . 41 "

Therefore,

150 violet-red flower color	(*Aa*)	occurred in	85	plants
white " "	(*a*)	" "	81	"
long stem " "	(*Bb*)	" "	87	"
short stem " "	(*b*)	" "	79	"

In this experiment, too, the proposed hypothesis finds adequate
155 confirmation.

Experiments on a small scale were also made on the traits of *pod shape, pod color,* and *flower position,* and the results obtained

were in full agreement: all combinations possible through union of
the different traits appeared when expected and in nearly equal numbers.
160 Thus experimentation also justifies the assumption *that pea hybrids form germinal and pollen cells that in their composition correspond in equal numbers to all the constant forms resulting from the combination of traits united through fertilization.*

Interpretive Comments

In this series of experiments Mendel has set out to test his theory of hybridization using traits of his peas other than the seed traits he has used thus far. This is a necessary next step in showing that his ideas apply to all the traits of peas that he selected and not just those of the seeds.

Let us call to mind again the assumptions of his theory. He has assumed (1) that hybrids form as many kinds of egg and pollen cells as there are possibilities for constant combinations when the hybrids reproduce by self-fertilization; (2) that these cells correspond in their internal compositions to those of the constant forms; (3) that the types of cells are produced in very nearly equal numbers; and (4) that in fertilization the egg and pollen cells combine at random in all possible combinations.

Here Mendel reports on an elegant series of experiments that he devised to provide another test of his theory of hybridization. He knew from previous experience that the choice of hybrids, as a source of egg and pollen cells, was of primary importance in determining the success of such experiments. In this particular case he thought that the simplest way to test his ideas was to cross two special hybrids involving only two traits, each one hybrid for one trait and constant for the other. Each of these two hybrids would have to give only two kinds of germ and pollen cells. If the trait patterns were reversed in the two hybrids, this would give four possible combinations in their progeny, which should be present in roughly equal proportions or numbers. This, in outline, is what Mendel did in this series of experiments. However, he skipped lightly over some of the steps he went through in explaining what he did, thinking perhaps that those who heard him speak or read his paper would be able to fill in the missing material. As it turns out, finding the missing pieces is not that easy for those of us who are not plant hybridizers of his stature. Having worked them out, we shall share them with you now.

Looking over his breeding stock Mendel selected three true-breeding lines, which he could combine in pairs to produce the two special hybrids he needed to test his ideas. One of these lines bred true for violet red flowers (*A*) and short stems (*b*). He symbolized this line as (*Ab*). Another was true-breeding for white flowers (*a*) and long stem (*B*). He symbolized this line as (*aB*). The third line was true-breeding for white flowers (*a*) and short stems (*b*). This one he symbolized as (*ab*). Note that, in the trait flower color, violet red is dominating and white is recessive, while in the trait stem length, the character long stem is dominating and short stem is recessive.

Mendel produced his first special hybrid by pollinating a plant that had violet red flowers and short stems with the pollen of a plant that had white flowers and short stems. Using Mendel's symbolism, the cross can be written $Ab \times ab \leftrightarrow Aab$. Note that Mendel mentions this cross in line 126, but does not fill in all the details.

He produced his second special hybrid by pollinating a plant that had white flowers and a long stem with a plant that had white flowers and a short stem. The cross can be written $aB \times ab \leftrightarrow aBb$. Again Mendel mentions this cross in lines 126 and 127, but does not give us any details. On the basis of his assumptions the first special hybrid would produce two possible types of germinal cells, *Ab* and *ab,* and the second would produce two possible types of pollen cells, *aB* and *ab,* as Mendel shows in line 131. If they were combined in fertilization in all possible combinations, the combinations and the possible color of the flowers on the plants grown from the seeds produced by the fertilizations should be as shown in lines 137–140. We expand this set in table 9.3 in order to make Mendel's reasoning clearer. Half the plants should have violet red flowers and half should have white flowers.

In the same way, we can follow the possible stem-length combinations (table 9.4). Again, half of the plants should have long stems and half should have short stems.

Looking at both traits, there should be four combinations produced as Mendel shows us in symbolic form in line 134. Of these, the combination *AaBb* would have violet red flowers and long stems; the combination *aBb* would have white flowers and long stems; the combination *Aab* would have violet red flowers and short stems; and the

Table 9.3. Predicted Flower Color from Second Special Cross

	Germinal cell	Pollen cell	Color trait combination	Predicted flower color
1	Ab	aB	Aa	violet red
2	Ab	ab	Aa	violet red
3	ab	aB	a	white
4	ab	ab	a	white

Table 9.4. Predicted Stem Length from Special Cross

	Germinal cell	Pollen cell	Stem length combination	Predicted stem length
1	Ab	aB	Bb	long
2	Ab	ab	b	short
3	ab	aB	Bb	long
4	ab	ab	b	short

combination *ab* would have white flowers and short stems. Each of these forms should be present in nearly equal numbers, or in proportional representation each form should be present as one-fourth of the total number.

Having made these predictions, Mendel proceeded to test them in the experiments he begins to report in line 141. Lines 144–148 summarize his results from the 166 plants that flowered. Lines 150–153 show that the individual characters had appeared in the proportions predicted from his theory. Here the experiments support Mendel's theory, as he states in lines 160–163.

Mendel's Text
Block D: Monohybrid and Dihybrid Experiments

The difference of forms among the progeny of hybrids, as well as the
165 ratios in which they are observed, find an adequate explanation in
the principle just deduced. The simplest case is given by the series
for *one pair of differing traits.* It is known that this series is
described by the expression: $A + 2Aa + a$, in which A and a signify
the forms with constant differing traits, and Aa the form hybrid for both.

170 The series contains four individuals in three different terms. In their production, pollen and germinal cells of form *A* and *a* participate, on the average, equally in fertilization; therefore each form manifests itself twice, since four individuals are produced. Participating in fertilization are thus:

175 Pollen cells $A + A + a + a$
Germinal cells $A + A + a + a$

It is entirely a matter of chance which of the two kinds of pollen combines with each single germinal cell. However, according to the laws of probability, in an average of many cases it will always happen that
180 every pollen form *A* and *a* will unite equally often with every germinal-cell form *A* and *a;* therefore, in fertilization, one of the two pollen cells *A* will meet a germinal cell *A,* the other a germinal cell *a,* and equally, one pollen cell *a* will become associated with a germinal cell *A,* the other with *a.*

185

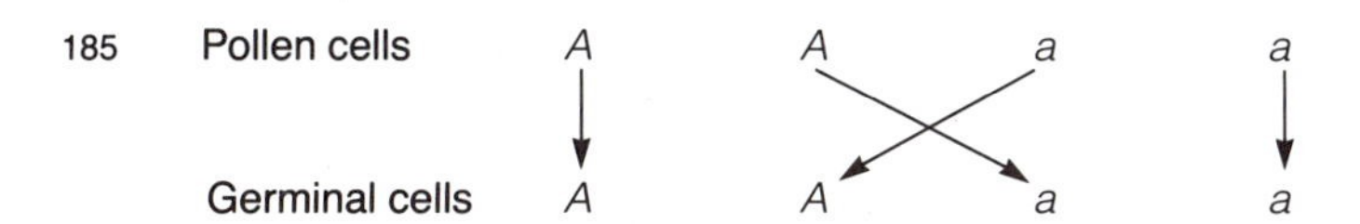

The result of fertilization can be visualized by writing the designations for associated germinal and pollen cells in the form of fractions, pollen cells above the line, germinal cells below. In the case
190 under discussion one obtains:

$$\frac{A}{A} + \frac{A}{a} + \frac{a}{A} + \frac{a}{a}$$

In the first and fourth terms germinal and pollen cells are alike, therefore the products of their association must be constant, namely *A* and *a;* in the second and third, however, a union of the two differing parental
195 traits takes place again, therefore the forms arising from such fertilizations are absolutely identical with the hybrid from which they derive. *Thus, repeated hybridization takes place.* The striking phenomenon, that hybrids are able to produce, in addition to the two parental types, progeny that resemble themselves is thus explained: $\frac{A}{A}$ and $\frac{a}{a}$ both give

200 the same association, Aa, since, as mentioned earlier, it makes no difference to the consequence of fertilization which of the two traits belongs to the pollen and which to the germinal cell. Therefore

$$\frac{A}{A} + \frac{A}{a} + \frac{a}{A} + \frac{a}{a} = A + 2Aa + a.$$

This represents the *average* course of self-fertilization of hybrids
205 when two differing traits are associated in them. In individual flowers and individual plants, however, the ratio in which the members of the series are formed may be subject to not insignificant deviations. Aside from the fact that the numbers in which both kinds of germinal cells occur in the ovary can be considered equal only on the average, it
210 remains purely a matter of chance which of the two kinds of pollen fertilizes each individual germinal cell. Therefore, isolated values must necessarily be subject to fluctuations, and even extreme cases are possible, as mentioned earlier in experiments on seed shape and albumen coloration. The true ratios can be given only by the mean calculated from
215 the sum of as many separate values as possible; the larger their number the more likely it is that mere chance effects will be eliminated.

The series for hybrids in which *two kinds of differing traits* are associated contains 16 individuals representing 9 different forms, namely: $AB + Ab + aB + ab + 2ABb + 2aBb + 2AaB + 2Aab + 4AaBb$. Among the
220 different traits of the parental plants, A, a and B, b, 4 constant combinations are possible; therefore the hybrid produces the 4 corresponding forms of germinal and pollen cells, AB, Ab, aB, ab, and each of these will fertilize or be fertilized 4 times on the average, since the series contain 16 individuals. Participating in fertilization
225 are thus:

Pollen cells: $AB + AB + AB + AB + Ab + Ab + Ab + Ab + aB + aB + aB + aB + ab + ab + ab + ab.$

Germinal cells: $AB + AB + AB + AB + Ab + Ab + Ab + Ab + aB + aB + aB + aB + ab + ab + ab + ab.$

230 In fertilization every pollen cell unites, on the average, equally often with each form of germinal cell; thus each of the 4 pollen cells AB once with each of the germinal cell forms AB, Ab, aB, ab. In precisely

the same manner the union of the remaining pollen cells of types *Ab, aB, ab* with all the other germinal cells take place. Thus one obtains:

235

$$\frac{AB}{AB} + \frac{AB}{Ab} + \frac{AB}{aB} + \frac{AB}{ab} + \frac{Ab}{AB} + \frac{Ab}{Ab} + \frac{Ab}{aB} + \frac{Ab}{ab}$$

$$\frac{\mathrm{aB}}{\mathrm{AB}} + \frac{\mathrm{aB}}{\mathrm{Ab}} + \frac{\mathrm{aB}}{\mathrm{aB}} + \frac{\mathrm{aB}}{\mathrm{ab}} + \frac{\mathrm{ab}}{\mathrm{AB}} + \frac{\mathrm{ab}}{\mathrm{Ab}} + \frac{\mathrm{ab}}{\mathrm{aB}} + \frac{\mathrm{ab}}{\mathrm{ab}}$$

or

$$AB + ABb + AaB + AaBb + ABb + Ab + AaBb + Aab + AaB + AaBb + aB + aBb + AaBb + Aab + aBb + ab$$

240

$$= AB + Ab + aB + ab + 2ABb + 2aBb + 2AaB + 2Aab + 4AaBb.$$

The series of hybrids in which *three kinds of differing traits* are combined can be explained in quite similar fashion. The hybrid produces 8 different forms of germinal and pollen cells: *ABC, ABc, AbC, Abc, aBC, aBc, abC, abc,* and again each pollen form unites once, on 245 the average, with each germinal-cell form.

The law of combination of differing traits according to which hybrid development proceeds thus finds its basis and explanantion in the proven statement that hybrids produce germinal and pollen cells that correspond in equal numbers to all the constant forms resulting from the 250 combination of traits united through fertilization.

Interpretive Comments

Having tested his theory by a series of specially prepared experiments, Mendel turned to the monohybrid, dihybrid, and trihybrid crosses he had done in order to find the empirical laws of formation of hybrids and their progeny.

In this concluding part of section 9 Mendel shows how the theoretical scheme he has developed, involving the production of certain kinds of reproductive cells of hybrids in certain numbers, can explain the results he had obtained in his experiments on progeny formation in monohybrids, dihybrids, and trihybrids. In lines 164 through 216 he shows how his scheme can explain why the progeny of monohybrids are always produced in the simple developmental series $A + 2Aa + a$. He demonstrates that it is necessary for the germinal and pollen cells to be produced in equal numbers and of the right kind, but that this is not enough. It is also necessary that the cells of the two kinds, *A* and *a*, should participate, on the average, equally in fertilization. Thus, in fertilization, the probability of an *A* pollen cell meeting an *a* germinal cell is the same as the probability it will meet an *A* germinal cell. The same is true for an *a* pollen cell. Mendel shows the pattern of these meetings in a diagram (lines 185–186). Then, in line 191 he shows the possible pairings of cells in fraction form, writing the pollen cells as numerators and the germinal cells as denominators. In the discussion that follows in lines 192 to 205 he demonstrates how this pattern of cell combinations explains not only the recovery of the true-breeding parental forms, but also how the hybrids can reproduce themselves even though they are not true-breeding. Mendel is then careful to point out (lines 205–216) the importance of using sufficiently large samples to avoid chance deviations that may occur in individual instances.

In line 217 he begins a parallel analysis of his dihybrid results. In lines 226–227 he sets down the pollen cells of a dihybrid in four groups, each with four cells, and in lines 228–229 the corresponding germinal cells. This parallels the arrangement for monohybrids given in line 191. Then, in lines 235–240 he writes out the pairings of pollen cells and germinal cells in fertilization in a fashion parallel to those in line 203 for the monohybrids. The summation of all these pairings results in the progeny series he had derived earlier by analysis of his experimental data.

When Mendel comes to the trihybrid analysis (lines 241–245) he does not write out the pairings of pollen and germinal cells taking place in fertilization. Instead he states that they would take place in a fashion similar to the dihybrids. In the concluding paragraph of this section he states (lines 246–250) what he feels he has established by this long series of experiments. That is, that he has developed a process or mechanism of the formation of reproductive cells and their combination in fertilization which explains the origin of "the law of combination of differing traits according to which hybrid development proceeds." He has succeeded in providing a theoretical scheme that is able to account for the empirically descriptive laws he derived from the analysis of his experimental data. However, he has done this with only a few traits of one plant, peas. The next question was whether or not the relationships he found in this one kind of plant would also hold true for other kinds. If the relationships did not hold true, would the same theoretical explanation of the relationship hold true? As we shall see, Mendel tried to find answers to these questions and reported the results of his efforts in section 10 of his paper.

Notes

1. What we have called Mendel's scheme was, in fact, a theory of hybridization. He probably felt that his work had laid the foundation for a precise quantitative science of hybrid formation. However, it was not recognized as such by his contemporaries. Even so, he retained his confidence. As he said to a friend a few years before his death, "My time will come." In 1900 his time did come but not in the way or in the form he had expected, for his theory of hybridization became the basis for a theory of heredity.

2. When Mendel wrote his paper in 1866, the idea that fertilization was the result of the fusion of two sex cells, one male and the other female, was in its infancy. Some of the greatest biologists of the time had a completely erroneous idea of the process. As late as 1840 they questioned sexuality in plants, believed that the embryo developed from the pollen tube itself after penetration into the pistil, and denied any union of the sperm with the egg. Mendel's biology teachers at the University of Vienna did not share the same opinion about fertilization. Fenzl, on the one hand, thought that the zygote (the first cell of the new organism) developed from the pollen cell. Unger, on the other hand, after working with mosses, thought that it was the result of the union of both male and female sex cells.

In 1855, one year before Mendel started his pea hybridizing experiments, the German botanist Nathaneal Pringsheim demonstrated that only one sperm is needed to fertilize the egg. He was able to do so by observing the union of a sperm cell with an egg cell in the freshwater alga *Oedogonium*. It seems that Mendel was well aware of Pringsheim's research and had it in mind when he developed his own view of fertilization.

§10

Experiments on Hybrids of Other Plant Species

As we have seen, Mendel's search for order in nature finally led him to his general law for *Pisum,* the combination series law. Having found the law, he proposed a hypothetical mechanism to explain why the law took the form of a combination series. This mechanism involved the formation of germinal and pollen cells and their union in fertilization. Great as these achievements were, they were still limited to a single kind of organism, peas. Thus, they were far short of that "generally applicable law of the formation and development of hybrids" for which Mendel was searching.

Recognizing the limited nature of what he had found, Mendel embarked on a further series of experiments with other kinds of plants to find out just how general his newly discovered law was. For example, if it proved to apply to beans, a closely related kind of plant, the search could then be extended to still other organisms. If it was partly successful the next step would be to modify the law and/or the explanatory mechanism in such a way as to adapt it to meet the new situation while saving as much of the original scheme as possible.

As you will see, this is precisely the problem Mendel confronted when he tried to extend his scheme for *Pisum.* For some of the characteristics, what he had found for peas held true for beans; for other characteristics, such as flower color, it did not.

Mendel's Text
Block A: Extending the Law for *Pisum* to Other Plants, First to Beans

1 The object of further experiments will be to determine whether the law of development discovered for *Pisum* is also valid for hybrids of

other plants. Several experiments were started quite recently for this purpose. I have completed two fairly small experiments with species of
5 *Phaseolus,* which might be mentioned here.

An experiment with *Phaseolus vulgaris* and *Phaseolus nanus* L. gave fully concordant results. *Ph. nanus,* in addition to a dwarf-like stem, had green smoothly arched pods; *Ph. vulgaris,* on the other hand, had a stem 10–12′ long and yellow pods, constricted at maturity. The
10 numerical relationship in which different forms occurred in individual generations were the same as in *Pisum.* The formation of constant associations also proceeded according to the law of simple combination of traits, exactly as in *Pisum.* Obtained were:

15 *Constant combination*	*Stem*	*Color of unripe pod*	*Shape of ripe pod*
1	long	green	arched
2	"	"	constricted
3	"	yellow	arched
4	"	"	constricted
20 5	short	green	arched
6	"	"	constricted
7	"	yellow	arched
8	"	"	constricted

The green pod color, arched pod shape, and tall stem were dominating
25 traits, as in *Pisum.*

Another experiment with two very different *Phaseolus* species was only partly successful. Serving as *seed plant* was *Ph. nanus* L., a very constant species with white blossoms in short racemes and small white seeds in straight, arched, smooth pods; as *pollen plant Ph.*
30 *multiflorus* W., with tall winding stem, crimson blossoms in very long racemes, rough sickle-like crooked pods and large seeds with black flecks and splashes on a peachblossom-red background.

The hybrid bore the greatest resemblance to the pollen plant, but the blossoms seemed less intensely colored. Its fertility was very limited;
35 from 17 plants that developed a total of many hundreds of blossoms, only 49 seeds were harvested. These were of medium size and bore a design similar to that of *Ph. multiflorus*; the background color also did not differ basically. In the following year they produced 44 plants of which only 31 reached the flowering stage. The traits of *Ph. nanus,* which all

became latent in the hybrid, reappeared in various combinations; their proportion to the dominating traits, however, fluctuated greatly because of the small number of experimental plants; but for some traits, like stem and pod shape, it was, as in *Pisum,* almost exactly 1:3.

Limited as the results of this experiment may be for the determination of ratios in which the various forms occurred, yet it provides a case of a *remarkable color change* in the blossoms and seeds of hybrids. It is known that in *Pisum* the traits of blossom and seed color appear unchanged in the first and in later generations, and that the offspring of hybrids carry exclusively one or the other of the two parental traits. The situation is different in the present experiment. True, the white flower and seed color of *Ph. nanus* appeared immediately in the first generation on one fairly fertile plant, but the remaining 30 plants developed flower colors that represented several gradations from crimson to pale violet. The coloration of the seed pod was no less varied than that of the flower. No plant could be considered fully fertile: some set no fruit at all; in others fruit was produced only by the last blossoms and did not have time to ripen. Well-formed seeds were harvested from only 15 plants. The greatest tendency toward infertility appeared in predominantly red-flowering forms; out of 16 such plants only 4 yielded ripe seeds. Three of these had a seed pattern similar to that of *Ph. multiflorus*, but a more or less pale background color, the fourth plant yielded only one seed, of plain brown coloration. Forms with preponderantly violet flower color had dark-brown, black-brown, and totally black seeds.

The experiment was continued for two more generations under equally unfavorable conditions, since even among the progeny of fairly fertile plants there were again some that were poorly fertile or completely sterile. No flower and seed colors other than those mentioned appeared. Forms receiving one or more of the recessive traits in the first generation remained constant in those traits without exception. Also, among the plants with violet blossoms and brown or black seeds, a few showed no further change in flower and seed color in the next generation, but the majority yielded, in addition to identical offspring, some with white flowers and similarly colored seed coats. Red-flowering plants remained so poorly fertile that nothing can be said with certainty about their further development.

Despite the many obstacles with which the observations had to contend, this experiment still establishes that development of hybrids follows the same law as in *Pisum* with respect to those traits concerned with the shape of the plant. Concerning the color traits, however, it seems difficult to find sufficient agreement. Besides the fact that a union of white and crimson coloration produces a whole range of colors from purple to pale violet and white, it is also striking that out of 31 flowering plants only one received the recessive trait of white coloration, while in *Pisum* this is true of every fourth plant on the average.

But these puzzling phenomena, too, could probably be explained by the law valid for *Pisum* if one might assume that in *Ph. multiflorus* the color of flowers and seeds is composed of two or more totally independent colors that behave individually exactly like any other constant trait in the plant. Were blossom color A composed of independent traits $A_1 + A_2 + \ldots$, which produce the overall impression of crimson coloration, then, through fertilization with the differing trait of white color a, hybrid associations $A_1a + A_2a + \ldots$ would have to be formed; and the situation with the corresponding coloration of the seed coat would be similar. According to the above assumption, each of these hybrid color combinations would be independent, and, therefore, would develop entirely independently from the rest. Then it is easily seen that from the combination of the individual series a complete color range should result. If, for instance, $A = A_1 + A_2$, then the series that correspond to hybrids A_1a and A_2a are

$$A_1 + 2A_1a + a,$$
$$A_2 + 2A_2a + a.$$

The terms of these series can enter into 9 different combinations, each of which represents the designation for another color:

$$\begin{array}{lll} 1\ A_1\ A_2 & 2\ A_1a\ A_2 & 1\ A_2\ \ a, \\ 2\ A_1\ A_2a & 4\ A_1a\ A_2a & 2\ A_2a\ a, \\ 1\ A_1\ a & 2\ A_1a\ a & 1\ a\ \ \ a \end{array}$$

The numbers preceding the individual combinations indicate how many plants of corresponding coloration belong to the series. Since their sum is 16, all colors are distributed over each 16 plants on the average, but, as the series itself shows, in unequal proportions.

If color development really occurred in this manner, then the above-mentioned case of white blossom and seed-coat color appearing only once
115 among 31 plants of the first generation would have an explanation. This coloration occurs only once in the series and, therefore, could be expressed only in every 16 plants, on the average; for three color traits once only even among 64 plants.

One must not forget, however, that the explanation attempted here
120 rests on a mere supposition, with nothing more to commend it than the very incomplete results of the experiment just discussed. It would be a worthwhile task, though, to follow color development in hybrids further by similar experiments, because it is probable that through this approach we can learn to understand the extraordinary diversity in the *coloration of*
125 *our ornamental flowers*.

Interpretive Comments

As we said in the introduction to this section, after discovering the law of development for *Pisum*, Mendel was anxious to find out whether or not the law was also valid for other plant species. So he turned his attention from peas to beans since both are members of the same family, the Leguminosae. The experiments he carried out with beans were not as extensive as those he carried out with peas (lines 1–5). Nevertheless, he was able to confirm in one experiment that the law was valid for at least two traits of beans, the color of the unripe pod and the shape of the ripe pod (lines 6–25). However, in a second experiment, when Mendel turned to another trait, flower color, the problem became more complex.

The parental plants in his second breeding experiment were *Phaseolus multiflorus* and *Ph. nanus*. These types of beans differ not only in the color of their flowers, but also in the color of the seeds, in the way the flower stalks are clustered, and in the shape of the pods. The last two traits behave in beans the same way as they do in peas: the hybrid plants showed the dominating characters for those two traits. Although Mendel does not tell us exactly this, it can be deduced from his observation that the hybrids resembled most closely the pollen plant (line 33). The pollen plant had most of the dominating characters (lines 30–32). In the case of the color of the flowers, the hybrid showed less intensity than that of the pollen parent (line 34), which had crimson blossoms (line 30). Therefore, this trait did not behave the way that flower color did in peas.

Unfortunately for Mendel, the hybrid plants that he obtained were not very productive. Although they produced a lot of flowers, they set only a few seeds (49 seeds from 17 plants, lines 35–36). Only 31 plants grew from these 49 seeds. Quantitative studies of stem length and pod shape confirmed that those traits behaved as they did in peas. Though the number of plants was small, Mendel found that the proportion of recessive characteristics to dominating ones was almost exactly 1:3 (line 43).

However, contrary to what he had found in peas, Mendel discovered that in beans the flowers, seeds, and pods showed a wide range of colors. In lines 44–64 he describes and discusses this wide range of colors and fertility in the offspring of the hybrids between *Phaseolus multiflorus* and *Ph. nanus*. It seems that there was a relationship between infertility and the degree of coloration among those plants (lines 58–59). Among the 31 offspring of the hybrid, only one had white flowers—when Mendel must have expected 8. The others had flowers ranging from "crimson to pale violet" (lines 53–54).

Confronted with such unexpected results, but still hopeful, Mendel continued this experiment for two more generations. Unfortunately, he again ran into the problem of low fertility (lines 67–68). Even so, he was able to come to several conclusions.

The first was that the complete range of variability of flower and seed color—the same color appears in the flower and the seed—existed among the first generation of offspring of the hybrids, since, in the next two generations, Mendel did not find any colors "other than those mentioned" (line 68). The second is that some of the plants were true-breeding for flower or seed color (lines 70–71), while others were still producing flowers of different colors (lines 73–74). Mendel made the following statement: "Forms receiving one or more of the recessive traits in the first generation remained constant in those traits without exception" (lines 69–70). It is not clear whether Mendel is referring to traits other than flower or seed color. But since these colors did not show the phenomenon of dominance or recessiveness, Mendel must have been referring to the shape of pod and stem, traits for which he believed that the law for *Pisum* was valid. This is the essence of what he states in lines 77–80.

Although Mendel could conclude that the law for *Pisum* was valid for traits in beans concerned "with the shape of the plant" (line 80), he could not draw the same conclusion for the color traits in his beans. After all, he found only one white flowered plant among the 31 colored ones (lines 84–86), when according to what he had found in peas, he should have found one out of 4. In addition, there was the problem that the flowers of 30 other plants differed widely in color. Mendel then proposed a hypothesis (lines 87–96) to explain these results. The hypothesis, which can be considered an extension of his law of development for *Pisum*, is as follows: in "*Ph. multiflorus* the color of flowers and seeds is composed of two or more totally independent colors that behave individually exactly like any other constant trait in the plant" (lines 88–91). He symbolized these *independent* colors by A_1, A_2, and so on. The most brilliant color, crimson, would be composed of all these independent colors. We have italicized the word "independent" above because it leads to the basic idea in the discussion that follows.

Suppose that a crimson-flowered bean were crossed to a white-flowered bean. Suppose, for simplicity's sake, that crimson flower color is composed of only two independent colors, A_1 and A_2, and that white flower color is a simple recessive character, denoted by a. In the formation of the hybrid, each of these independent colors, A_1 and A_2, would associate with a, forming character pairs. According to Mendel, the color of the flowers of the hybrid of these two bean varieties would be the result of the association of these character pairs. Mendel assumed that these character pairs were transmitted to the offspring of the hybrid independently of each other—the same way, let us say, as the character pairs affecting stem size or color of cotyledons in peas. If this assumption was correct, we would have three classes of hybrid offspring for both character pairs. The series that would correspond to hybrids A_1a and A_2a would be

$$A_1 + 2A_1a + a \text{ and } A_2 + 2A_2a + a$$

The terms of these series (lines 102–103) can enter into 9 different combinations, each of which represents the designation for another color (lines 104 and 105). Mendel gives us these 9 different combinations on lines 106–108. Note that these 9 different combinations of color are in the ratio 1:1:1:1:2:2:2:2:4 or 1:2:1:2:4:2:1:2:1, the same ratio that Mendel found in his dihybrid cross. In the present case this yields one plant with white flowers to 15 plants with colored flowers. Mendel found a plant with white flowers only once in the 31 plants he grew, as he notes in lines 114–115. This is virtually a 2 to 1 difference, yet Mendel makes no comment on this discrepancy. Such a discrepancy could be the result of the small sample of plants Mendel worked with; a small sample is never representative of a true ratio, as Mendel was well aware. On the other hand, it could be an indication that Mendel's scheme was incorrect.

Note that the last combination (line 108) is written *aa*, which is the one that gives white color. This is the only place in his paper that Mendel represents a character with two small letters. As you remember, he has always designated a character by one letter except in the case of a hybrid. This use of *aa* has led to controversy as to what Mendel meant. Some have seen it as evidence that Mendel was thinking about units of inheritance. But we believe that the reason Mendel wrote *aa* and not *a* is that he had to complete his combination series. For all his efforts here, it is evident (lines 119–121) that Mendel was not satisfied with his results. He insists that the explanation he gave was very tentative, based on a "mere supposition." He was very cautious, for he knew that his data were very incomplete since he had too few plants to work with and because of their low fertility. He suggests that further experiments should be made "to understand the extraordinary diversity in the *coloration of ornamental flowers*" (lines 124–125). He was perfectly right to be cautious because his hypothesis was later shown to be incorrect. Color formation in flowers is very complex.[1]

Notes

1. Since Mendel's time we have found that the inheritance of pigmentation in bean flowers, as in any other plant, is highly complex. It is now known that differences in flower color in general are determined by the ratio of two different types of pigments, called **anthocyanins** and **flavones,** which are dissolved in the sap of plant cells. The color of the flower depends not only on the concentration of these pigments, but also on the acidity of the cell sap. When the cell sap is acid in reaction, the color is often red, and when the sap is alkaline, it is usually blue. This can be demonstrated in the laboratory by placing cut blue flowers in a weak acid solution. After enough time has elapsed to permit the acid to penetrate up the stem to the flowers, they turn red. On the other hand, cut pink or red flowers will often turn blue when weak ammonia solution is used. Genetically speaking, we now know there is a basic gene in

flowering plants that controls the presence or absence of anthocyanins at the very beginning of the synthesis of those pigments. There are two other major genes necessary for their synthesis and other, minor, genes that modify the expression of these two major genes.

Mendel's Text
Block B: Remarks on Variability of Plants

Up to now, hardly more is known with certainty than that the flower
color in most ornamental plants is an extremely variable trait. The
opinion has often been expressed that, through cultivation, species
stability is greatly upset or entirely shattered, and there is a strong
130 inclination to describe the development of cultivated forms as devoid of
rules and subject to chance; usually the coloration of ornamental plants
is pointed out as a model of instability. However, it is not clear why
mere transplantation into garden soil should have such thorough and
persistent revolution in the plant organism as its consequence. No one
135 would seriously want to maintain that plant development in the wild and
in garden beds was governed by different laws. Here as well as there
changes in the type must appear when living conditions are changed and
when a species has the ability to adapt itself to the new environment.
Granted willingly that cultivation favors the formation of new varieties
140 and that by the hand of man many an alteration has been preserved which
would have perished in nature, but nothing justifies the assumption that
the tendency to form varieties is so extraordinarily increased that
species soon lose all stability and their progeny diverge into an
infinite number of extremely variable forms. If the change in living
145 conditions were the sole cause of variability one could expect that those
cultivated plants that have been grown through centuries under almost
identical conditions should have regained stability. This is known not to
be the case, for it is precisely among them that not only the most
different but also the most variable forms are found. Only Leguminosae,
150 such as *Pisum*, *Phaseolus*, and *Lens*, whose organs of fertilization are
protected by the keel, represent notable exceptions. During more than
1000 years of cultivation under the most diversified conditions, numerous
varieties have arisen, yet these maintain stability under constant living
conditions, just as do species growing wild.

155 It remains more than probable that a factor that so far has received
little attention is involved in the variability of cultivated plants.

Various experiences force us to accept the opinion that our cultivated plants, with few exceptions, are *members of different hybrid series* whose development along regular lines is altered and retarded by frequent
160 intraspecific crosses. It should not be overlooked that cultivated plants are usually raised in fairly large numbers in close proximity to each other, a condition most favorable for reciprocal fertilization among the varieties present and between the species themselves. The likelihood that this opinion is correct is supported by the fact that among the large
165 array of variable forms one finds always some single ones that remain constant for one or the other trait if all extraneous influence is carefully excluded. These forms develop exactly like certain members of the composite hybrid series. Even with respect to the most sensitive of all traits, that of color, it cannot escape careful observation that a
170 tendency to variability exists in the individual forms to a very different degree. Among plants originating from a *single* spontaneous fertilization there are frequently some whose progeny diverge widely in the type and disposition of colors, while others produce forms that deviate little, and, if the number of plants is fairly large, some are
175 encountered that transmit the color of their flowers unchanged to their progeny. Cultivated *Dianthus* species are an instructive example of this. A white-flowering specimen of *Dianthus caryophyllus*, itself derived from a white-flowered variety, was isolated in a greenhouse during the flowering period; its numerous seeds grew into plants with flowers of
180 exactly the same shade of white. A similar result was obtained from a red, slightly violet glistening sport and from a white one with red stripes. On the other hand, many others, protected in the same manner, produced more or less differently colored and patterned progeny.

Anyone surveying the shades of color that appear in ornamental plants
185 as a result of like fertilization cannot easily escape the conviction that here, too, development proceeds according to a certain law which possibly finds its expression through the *combination of several independent color traits*.

Interpretive Comments

Though his studies of coloration in flowers were sketchy, Mendel doubted that color in ornamentals varied because of their transfer from the wild to gardens. He could not accept the idea that the development of cultivated forms was different from that of wild forms. For him, they were subject to the same laws. After ending his studies with peas, he was even more convinced than he was at the beginning of his experiments that there was order in the biological world and that there were laws governing the development of organisms. And so he asked why the "mere transplantation into garden soil" should change the makeup of plants (line 133). In lines 134–136 Mendel is arguing that, wherever the plant is grown, its development is subject to the same laws. This idea, a fundamental one in Mendel's mind, permitted him to carry out his experiments. The next sentence (lines 136–138), however, is far from being clear. Mendel writes: "Here as well as there changes in the type must appear when living conditions are changed and when a species has the ability to adapt itself to the new environment." Mendel was not specific as to the nature of these changes. Are these changes permanent, that is, are they carried from one generation to the next? Or are they temporary, that is, do they disappear after one generation? Is Mendel talking about a temporary adaptation or about a permanent adaptation? We do not know.[1]

It is true that we have been able to select and propagate varieties that otherwise would have perished in the wild. But why should plants under cultivation have a greater tendency to "form varieties" and be more variable than they are in the wild? To counter this idea, Mendel offers the fact that many cultivated forms that should have remained stable when grown under the same environment had given rise to "the most different but also the most variable forms" (lines 148–149). On the other hand, the family of Leguminosae, to which peas, beans, and lentils belong (lines 149–150), has also produced, under the most diversified cultivation practices, a large number of varieties (possibly by human intervention). Yet under "constant living conditions" (lines 153–154), that is, left to themselves, they are as stable as wild species. The reason for this stability in legumes is that their flowers remain closed during pollination. This gave a clue to Mendel that it was not cultivation per se that rendered plants more variable, but that hybridization between different varieties or types might happen more freely under cultivation than in the wild. Cultivated plants are usually raised in fairly large numbers and in close proximity to other types of plants, thus permitting cross-fertilization. Among the offspring of the resulting hybrids, some will be true-breeding, "constant for one or the other trait" (line 166). This was a conclusion Mendel had reached from his pea crosses and also from some of his bean crosses. He found that there were different degrees of variability among the many forms of the offspring of the hybrid. Some were very variable, some slightly variable, and some not at all variable.

As an example of this, Mendel returned to the problem of flower color. He wrote: "Among plants originating from a single spontaneous fertilization there are frequently some whose progeny diverge widely in the type and disposition of colors, while others produce forms that deviate little" (lines 171–174). By "*single* spontaneous fertilization" Mendel meant that one flower was naturally cross-fertilized with the pollen of another type of plant. If there were numerous plants among the progeny of such a cross, some of them would transmit their flower color unchanged to their own progeny (that is, they would be true-breeding for flower color). Most of the others would not be true-breeding, so flower color among their own progeny would be highly variable. To illustrate both cases, Mendel gave us examples of cultivated carnations.

In the last paragraph of this section, Mendel reaffirms his conviction that the development of flower color proceeds "according to a certain law," the same law that he had discovered for other traits. However, here, instead of having different traits, he deals with a compound trait, color, made up of "*several independent* [single] *color traits*." In other words, Mendel has extended his idea of combination series to a complex trait.

Notes

1. When Mendel wrote this sentence, his contemporaries did not distinguish between hereditary and environmental changes. The incorrect idea that characteristics *acquired* during one's lifetime could be transmitted to the next generation was still prevalent.

§11

Concluding Remarks

At this point Mendel has completed the exposition of his experimental work. He has shown how he accumulated quantitative information about the behavior of a graded series of pea hybrids ranging from the simplest possible ones to very complex ones. At each level of complexity of the hybrids he found concordant behaviors, which enabled him to formulate a series of quantitative generalizations, or empirical natural laws. In addition, he has developed a master law, the combination series law, that included all of the earlier empirical laws. In all of this he has confined himself to generalizations and laws derived from experimental data and which described those data but did not explain why the generalizations and laws took the form they did. This type of program was consistent with his view of himself as basically an empirical worker and not a theoretician. He designed and carried out his experiments as we would expect a person to do who was well trained in physics and mathematics as well as in botany and plant breeding.

But, true to his training in the physical sciences, he felt required to do two more things: first, to provide a hypothetical mechanism to account for the laws he found; second, to test the generality of the law for *Pisum* by attempting to apply it to other plants to find out whether it would work successfully with them or fail in whole or in part. We have followed him through all these stages of development as he described them to his audience.

What we have come to now are his final reflections on his work. As you will see, he is concerned with how his work related to that of his famous precursors, the correctness of whose work he trusted almost completely—almost, but not quite, as he points out in his observations on the importance of the size of samples they may have used, and not entirely when they reported the existence of hybrids that bred true, something he had never encountered in any of his many experiments (and none were found in experiments by other, later hybridizers).

Yet he honored their assertions so much that he spends considerable time and energy in proposing a hypothetical explanation of how his combinations of germinal and pollen cells might differ in true-breeding and in variable hybrids. Even though it is an ingenious speculation—which is wrong—he does not find it really convincing.

He turns from this to another problem, a residue from experiments made by his precursors who were trying to convert one species of plant into another, something that appeared possible given their ideas of what species were. Mendel is able to show, using his experimental results with peas, that they had obtained their results the hard way, and he points out a much simpler way of getting the same results. He concludes his remarks with some comments on the relation of these experiments to the idea that species could be changed only within narrow limits. These remarks have been read by some as evidence of Mendel's ideas about evolution versus special creation. However, since we now know that species cannot be converted by the methods these earlier breeders were using, this is not a very useful discussion.

Mendel's Text
Block A: General Comments on Hybrids

1 A comparison of the observations made on *Pisum* with the experimental
results obtained by Kölreuter and Gärtner, the two authorities in this field,
cannot fail to be of interest. Both concur in the opinion that, in
external appearance, hybrids either maintain a form intermediate between
5 the parental strains or they approach the type of one or the other,
sometimes being barely distinguishable from them. Various forms that
diverge from the normal type usually arise from the seeds of the hybrids
that were fertilized by their own pollen. As a rule the majority of
individuals produced from such a fertilization maintain the form of the
10 hybrid, a few become more like the seed plant, and an occasional
individual very nearly matches the pollen plant. This, however, is not
valid for all hybrids without exception. Among the offspring of certain
individuals some are more like one original stock plant, some more like
the other, or they all tend more to one side than the other; but those
15 from a few remain *exactly like the hybrid* and propagate unchanged. The
hybrids of varieties behave like species hybrids, but possess a still

greater inconstancy and a more pronounced tendency to revert to the original forms.

With respect to the *features* of hybrids and their regular *development,* consistency with the observations made on *Pisum* is unmistakable. This is not so in the exceptional cases mentioned. Gärtner himself admits that precise determination of whether a form bears a greater resemblance to one or the other of the two parental types often presents great difficulties, since much depends on the subjective viewpoint of the observer. And there is yet another circumstance that could contribute to making the results variable and uncertain in spite of the most careful observation and discrimination. For the most part plants which are considered to be good species and that differ in a rather large number of traits were used in the experiments. When one is dealing in a general way with degrees of similarity, then account must be taken not only of the traits that stand out sharply, but also of those that are often difficult to put into words, yet, as everyone familiar with plants knows, are sufficiently pronounced to give such forms the appearance of a stranger. If it is assumed that development of hybrids follows the law valid for *Pisum,* then the series obtained in each separate experiment must comprise very many forms, because the number of terms is known to increase with the number of differing traits as a power of three. Thus with a relatively small number of experimental plants the result could be only approximately correct and occasionally could deviate not inconsiderably. If, for instance, the two original stocks differed in 7 traits, and if 100 to 200 plants were raised from the seeds of their hybrids for an evaluation of the offsprings' degree of relationship, we can easily understand how uncertain such judgment must be, since the series for 7 differing traits contains 16,384 individuals appearing in 2187 different forms. Sometimes one relationship, sometimes another, would assert itself more strongly, depending on whether the observer found, by chance, a larger number of this or of that form.

Furthermore, when the differing traits include *dominating* ones that are passed on to the hybrid totally or almost totally unchanged, then the one of the two parental types having the larger number of dominating traits must always be the more prominent among the members of the series. In the experiment with three differing traits in *Pisum* described earlier, all of the dominating characters belonged to the seed plant. Although the members of the series tend equally toward both original parents in their internal makeup, the appearance of the seed plant was so preponderant in this experiment that 54 plants out of every 64 in the first generation looked exactly like it, or differed from it in only one trait. One sees how risky it can sometimes be to draw conclusions about the internal kinship of hybrids from their external similarity.

60 Gärtner mentions that in cases where development was regular the two parental types themselves were not represented among the offspring of the hybrids, only occasional individuals closely approximating them. Indeed, it cannot be otherwise in very extensive series. For 7 differing traits, for instance, each parental form occurs only once in more than 65 16,000 offspring of the hybrid. Therefore there is not much likelihood of finding them among a small number of experimental plants, yet, with a reasonable degree of probability, one may count on the appearance of a few forms that approximate those in the series.

We encounter an *essential difference* in those hybrids that remain 70 constant in their progeny and propagate like pure strains. According to Gärtner these include the *highly fertile* hybrids *Aquilegia atropurpurea-canadensis, Lavatera pseudolbia-thuringiaca, Geum urbano-rivale,* and some *Dianthus* hybrids; according to Wichura it includes the hybrids of willow species. This feature is of particular importance to the evolutionary 75 history of plants, because constant hybrids attain the status of *new species.* The correctness of these observations is vouched for by eminent observers and cannot be doubted. Gärtner had the opportunity of following the *Dianthus Armeria-deltoides* to its tenth generation, since that plant propagated itself regularly in the garden.

80 It was proven experimentally that in *Pisum* hybrids form *different kinds* of germinal and pollen cells and that this is the reason for the variability of their offspring. For other hybrids whose offspring behave similarly, we may assume the same cause; on the other hand, it seems permissible to assume that the germ cells of those that remain constant 85 are identical, and also like the primordial cell of the hybrid. According to the opinion of famous physiologists, propagation in phanerogams is initiated by the union of one germinal and one pollen cell into one single cell,[1] which is able to develop into an independent organism through incorporation of matter and the formation of new cells.

1. It is presumably beyond doubt that in *Pisum* a complete union of elements from both fertilizing cells has to take place for the formation of a new embryo. How else could one explain that both parental types recur in equal numbers and with all their characteristics in the offspring of hybrids? If the influence of the germinal cell on the pollen cell were only external, if it merely played the role of a foster mother, then the

90 This development proceeds in accord with a constant law based on the material composition and arrangement of the elements that attained a viable union in the cell. When the reproductive cells are of the same kind and like the primordial cell of the mother, development of the new individual is governed by the same law that is valid for the mother
95 plant. When a germinal cell is successfully combined with a *dissimilar* pollen cell, we have to assume that some compromise takes place between those elements of both cells that cause their differences. The resulting mediating cell becomes the basis of the hybrid organism whose development must necessarily proceed in accord with a law different from
100 that for each of the two parental types. If the compromise be considered complete, in the sense that the hybrid embryo is made up of cells of like kind in which the differences are *entirely and permamently mediated,* then a further consequence would be that the hybrid would remain as constant in its progeny as any other stable plant variety. The
105 reproductive cells formed in its ovary and anthers are all the same and like the mediating cells from which they derive.

One could perhaps assume that in those hybrids whose offspring are *variable* a compromise takes place between the differing elements of the germinal and the pollen cell great enough to permit the formation of a
110 cell that becomes the basis for the hybrid, but that this balance between the antagonistic elements is only temporary and does not extend beyond the lifetime of the hybrid plant. Since no changes in its characteristics can be noticed throughout the entire vegetative period, we must further conclude that the differing elements succeed in escaping
115 from the enforced association only at the stage at which the reproductive cells develop. In the formation of these cells all elements present participate in completely free and uniform fashion, and only those that differ separate from each other. In this manner the production of as many kinds of germinal and pollen cells would be
120 possible as there are combinations of potentially formative elements.

This attempt to relate the important difference in the development of hybrids to a *permanent or temporary association* of differing cell

---

outcome of each artificial fertilization would have to be that the resulting hybrid resembled the pollen plant exclusively or very closely. Experiments have in no way confirmed this up to now. A thorough proof for complete union of the content of both cells presumably lies in the universally confirmed experience that it is immaterial to the form of the hybrid which of the parental type was the seed or pollen plant.

elements can, of course, be of value only as a hypothesis which, for
lack of well-substantiated data, still leaves some latitude. Some
125 justification for the opinions expressed lies in the proof cited here
that in *Pisum* the behavior of a pair of differing traits in hybrid union
is independent of any other differences between the two parental plants
and that, furthermore, the hybrid produces as many kinds of germinal and
pollen cells as there are possible constant combination forms. The
130 distinguishing traits of two plants can, after all, be caused only by
differences in the composition and grouping of the elements existing in
dynamic interaction in their primordial cells.

Yet even the validity of the laws proposed for *Pisum* needs
confirmation, and a repetition of at least the more important
135 experiments is therefore desirable: for instance, the one on the
composition of hybrid fertilizing cells. An individual observer can
easily overlook a distinguishing point that seems unimportant in the
beginning but can grow to such proportions that it may not be neglected
in the final analysis. Whether variable hybrids of other plant species
140 show complete agreement in behavior also remains to be decided
experimentally; one might assume, however, that no basic difference
could exist in important matters since *unity* in the plan of development of
organic life is beyond doubt.

Interpretive Comments

In his concluding remarks Mendel compared his hybrid results with those of his predecessors, in particular Kölreuter and Gärtner. We should note that in this section Mendel speculated far more than in any of the other sections. His writing became more abstract, less factual. As a result, there are some places where it is not clear what he meant to convey.

In the first paragraph of this section, Mendel told us that both Kölreuter and Gärtner reported that, with many plants, their hybrids were either intermediate between the parental plants in their general appearance or they were more or less like one of the parents. Mendel did not make any comment on this, although he should have, since the statements made by Kölreuter and Gärtner were very vague and not at all illuminating. The reason for this vagueness is that they, like most of the hybridizers before them and some after them, compared the appearance of their hybrids with the appearance of the parents in toto. To judge a plant this way is very imprecise. A plant has a very large number of traits, and for some traits the hybrid might look like one parent; for others it might look like the other parent; and for others it might not look like either parent. This is what Mendel found out when he restricted his study first to one trait, then two, then three. Studying well-defined individual traits instead of the whole plant was fundamental to Mendel's success. By working in this way he was able to simplify and solve a problem that had confronted hybridizers for a long time, namely, the tendency of hybrids to revert to the parental form.

Mendel started to offer a solution to this problem (lines 34–35) by assuming that he could generalize his results with peas to other plants. He wrote: "If it is assumed that development of hybrids follows the law valid for *Pisum* . . ." We know from observation that a plant has a very large number of traits. As the number of traits that one considers is increased, the number of forms among the offspring of a hybrid increases, as Mendel demonstrated, as a power of 3. Mendel discusses this point on lines 36 and 37. If n designates the number of characteristic differences in the two parental plants, then 3^n is the number of terms in the combination series and 4^n the number of individuals that belong to the series (see table 8.1). In the example that Mendel gives, the two original stocks differ in seven traits; hence the number of individuals in the series will be $4^7 = 16{,}384$, appearing in $3^7 = 2{,}187$ forms. Therefore, if only 200 plants were to be raised from the seeds of the hybrid, it would be impossible to have all the possible forms of the offspring of the hybrid represented. In other words, one can in fact expect 3^n different types out of 4^n individuals among the offspring of the hybrid. From these 4^n individuals, only 2^n are constant (propagate unchanged) in the case of self-pollination, whereas only one of the 4^n may be expected to be entirely identical to either of the two original parents. Mendel concluded that this showed how difficult it was to draw conclusions from the analysis of some 100 to 200 individuals (lines 46–47).

Another reason for the difficulty encountered by Mendel's predecessors was that, for traits showing the dominating characters, there will be a preponderance of dominating forms among the offspring of the hybrid. Mendel explains this in lines 48 to 59. If, let us say, the seed parent is the one with the most dominating characteristics, then the offspring of the hybrid will, in general, have the appearance of the seed parent. On the other hand, if it is the pollen parent that has the most dominating characteristics, the offspring of the hybrid will, in general, have the appearance of the pollen parent. Thus, Mendel was completely correct when he said, "One sees how risky it can sometimes be to draw conclusions about the internal kinship of hybrids from their external similarity" (lines 57–59). This had first become clear to him when he allowed his monohybrids to reproduce by self-fertilization.

The small number of offspring that can be grown from a single hybrid plant also explains why, as Gärtner had observed, in many crosses the two parental types are not represented among the offspring. Mendel explained this mathematically (lines 60–68). We shall now develop his idea in some detail. Let us take two parents that differ in only one trait. We have seen previously that among the offspring of the hybrids of both parents, each parental form will appear once out of every four offspring. If the two parents differ in two traits, each parental form will appear once among 16 offspring. If the parents differ in three traits, each parental form will appear among 64 offspring. Taking Mendel's example, if the two parents differ in seven traits, the number of individuals in the series will be 16,384. Among those, each parental form will appear only once. Therefore, as Mendel points out, "there is not much likelihood of finding them among a small number of experimental plants" (lines 65–66). That being the case, it is no wonder that his predecessors could neither generalize nor deduce much from their observations.

In the next paragraph (lines 69–79) Mendel introduces the idea that there are two different types of hybrids, those that are constant (true-breeding) and those that are not. We should note that Mendel had not found any of his own hybrids to be true-breeding. The constant hybrids that he writes about were those of other workers, in particular those of Gärtner and Wichura. Although their results differed from his own, he took their observations at face value.[1] However, since Mendel's time we have discovered that Wichura's hybrids were not true-breeding. This might also have been true of Gärtner's hybrids, some of which degenerated in various ways (Olby 1984, 158).

In subsequent paragraphs (lines 80–139) Mendel attempts to explain how certain hybrids could be variable, that is, produce several kinds of offspring, while others could be nonvariable (true-breeding). He felt that he had proved (lines 80–82) that his pea hybrids, which were variable, formed different kinds of egg and pollen cells. Their random combination in fertilization produced the different kinds of progeny he had observed.

He explains the production of constant progeny from true-breeding plants by assuming that they produced pollen and egg cells that had the same composition. Therefore, nonvariable, or true-breeding, hybrids, *if they existed,* would produce egg cells and pollen cells having the same composition as "the primordial cell of the hybrid" (line 85).

Mendel could not know that what he had found for peas was also true for most plants. So, wisely, he suggested that his results with peas should be repeated (lines 133–135), though he was inclined to believe that fundamentally all plants should behave the same way since "*unity* in the plan of development of organic life is beyond doubt" (lines 142–143).

On lines 95–97 (block A of this section) Mendel made the following statement: "When a germinal cell is successfully combined with a *dissimilar* pollen cell, we have to assume that some compromise takes place between those *elements* of both cells that cause their differences."[2] This statement raises several important questions. First, what is the nature of this compromise? Mendel does not give us any clue but states clearly that the compromise is "only temporary and does not extend beyond the lifetime of the hybrid plant" (lines 111–112). Mendel wrote that the antagonistic *elements* escape from this compromise (forced association) only in the formation of the reproductive cells. This points to a second important question: what did Mendel mean by the word *element?*

Unfortunately, Mendel never defined the nature of these elements or the nature of the compromise between the cell elements of dissimilar cells. According to him, there were two traits in a hybrid, the dominating and the recessive one. The dominating was expressed, the recessive was not. Mendel realized that the compromise he suggested (lines 96–98) could not occur between the traits themselves because he (correctly) believed that traits are the expression of something in the cells. What did he think it was in the cells that determined these traits? He did not tell us, and here lies a weakness in his explanatory scheme. So long as he did not have a clear idea of the nature of the elements, he could not conceive the nature of the compromise. Years later, the nature of these determiners was discovered and the nature of the compromise understood. However, the discovery and the understanding did not come until the 1960s, when geneticists and biochemists working together were able to understand the functions of genes.

Mendel's Text
Block B: Transformation of One Species into Another

Finally, the experiments performed by Kölreuter, Gärtner, and
145 others on *transformation of one species into another by artificial*
fertilization deserve special mention. Particular importance was
attached to these experiments; Gärtner counts them as among the "most
difficult in hybrid production."

When species *A* was to be transformed into *B*, the two were combined
150 by fertilization and the resulting hybrids once more fertilized with
pollen from *B;* from among their various descendants those closest to
species *B* were then chosen and repeatedly fertilized by pollen from *B*,
and so on, until finally a form that was like *B* and remained constant in
its progeny was obtained. Thus species *A* was transformed into the other
155 species, *B*. Gärtner himself has carried out 30 experiments of this kind
with plants from genera *Aquilegia, Dianthus, Geum, Lavatera, Lychnis,*
Malva, Nicotinia, and *Oenothera.* The length of time needed for
transformation was not the same with all species. Although three
successive fertilizations were sufficient for some, with others
160 fertilizations had to be repeated five to six times; even with the same
species fluctuations were observed in different experiments. Gärtner
ascribes these differences to the circumstance that "the characteristic
force toward change and transformation of the maternal type that a
species exerts in reproduction is very different in different plants,
165 and consequently the length of time required for one species to become
transformed into another, and the number of generations it takes, must
also be different; transformation is accomplished after more generations
in some species, after fewer in others." The same observer notes further
"that in the process of transformation much depends on which type and
170 which individual was chosen for further transformation."

If one may assume that the development of forms proceeded in
these experiments in a manner similar to that in *Pisum*, then the entire process
of transformation would have a rather simple explanation. The
hybrid produces as many kinds of germinal cells as there are constant
175 combinations made possible by the traits associated within the hybrid,
and one of these is always just like the fertilizing pollen cells. Thus
there is the possibility that in such experiments a constant form
identical to the pollen parent will result from the second fertilization.

Whether one is actually obtained depends on the number of
180 plants in each experiment as well as on the number of differing traits that
were united by the fertilization. Let us assume, for example, that
the plants chosen for the experiment differ in three traits and that
species *ABC* is to be transformed into species *abc* by repeated
fertilization with pollen from the latter. The hybrid resulting from the
185 first fertilization forms 8 different kinds of germinal cells, namely:

ABC, ABc, AbC, aBC, Abc, aBc, abC, abc.

In the second year of the experiment these are again combined
with pollen cells *abc* and one obtains the series:

AaBbCc + *AaBbc* + *AabCc* + *aBbCc* + *Aabc* + *aBbc* + *abCc* + *abc.*

190 Since the form *abc* occurs once in the series of 8 terms there is
little likelihood that it would be missing among the experimental
plants, even if only a fairly small number were raised, and
transformation would thus be complete after two fertilizations. If, by
chance, no transformation was obtained, fertilization would have to be
195 repeated on one of the closest related combinations, *Aabc, aBbc, abCc.*
It becomes obvious that the *smaller the number of experimental plants
and the larger the number of differing traits* in the two parental
species the longer an experiment of this kind will last, and that
furthermore, a delay of one or even two generations could easily occur
200 with these same species, which is what Gärtner has observed. The
transformation of widely divergent species cannot be completed before
the fifth or sixth experimental year because the number of different
germinal cells formed in the hybrid increases with the number of
differing traits as a power of two.

205 Gärtner found by repeated experiments that the *reciprocal* period of
transformation varies for some species, so that quite frequently
species *A* can be transformed into species *B* a generation earlier than
species *B* into species *A.* From this he deduces that Kölreuter's
opinion that "the two natures in hybrids are in perfect equilibrium" is
210 not entirely tenable. It seems, however, that Kölreuter does not
deserve this reproach but rather that Gärtner has overlooked an

important point, to which he himself draws attention elsewhere, namely, that it "depends on which individual is chosen for further transformation." Experiments set up for this purpose with two *Pisum* species indicate that in the selection of individuals best suited for the purpose of further fertilization it could make a great difference which of the two species is to be transformed into the other. The two experimental plants differed in five traits; those of species *A* were all dominating, those of species *B* were all recessive. To effect mutual transformation *A* was fertilized with pollen from *B* and *B* with pollen from *A,* and the same procedure was repeated on both hybrids in the following year. In the first experiment, $\frac{B}{A}$, 87 plants *in the 32 possible forms* were available in the third experimental year from which to choose individuals for further fertilization; in the second experiment, $\frac{A}{B}$, the external appearance of all 73 plants obtained completely *coincided with that of the pollen plant,* although their internal constitution must have been just as varied as the forms from the other experiment. Intentional selection was therefore possible only in the first experiment; in the second one a few plants had to be chosen purely at random. Only a few flowers of the latter were fertilized with pollen from *A,* the rest were allowed to self-fertilize. Among each five plants used for fertilization in the two experiments, the next year's culture showed the following agreement with the pollen plant:

First experiment	*Second Experiment*	
2 plants	———	In all traits
3 "	———	" 4 "
———	2 plants	" 3 "
———	2 "	" 2 "
———	1 plant	" 1 trait

In the first experiment transformation was thus completed; in the second, which was not continued, two more fertilizations would probably have been necessary.

Though it is infrequently true that the dominating traits belong exclusively to one or the other parental plant, it will always make a difference *which* of the two possesses them in larger number. When the pollen plant has the majority of dominating traits, the choice of forms

for further fertilization will afford a lesser degree of certainty than
in the opposite case. A delay in the length of time needed for
250 transformation will be the consequence if the experiment be considered
complete only when a form is obtained that not only resembles the pollen
plant in appearance but, like it, remains constant in its progeny.

The success of transformation experiments led Gärtner to disagree
with those scientists who contest the stability of plant species and
255 assume continuous evolution of plant forms. In the complete
transformation of one species into another he finds unequivocal proof
that a species has fixed limits beyond which it cannot change. Although
this opinion cannot be adjudged unconditionally valid, considerable
confirmation of the earlier expressed conjecture on the variability of
260 cultivated plants is to be found in the experiments performed by
Gärtner.

Among the experimental species were cultivated forms such as
Aquilegia atropurpurea and *canadensis, Dianthus Carophylus chinensis*
and *japonicus, Nicotiania rustica,* and *paniculata,* and these, too, lost
265 none of their stability after 4 to 5 repetions of hybrid association.

Interpretive Comments

In the rest of his concluding remarks Mendel discussed how some of the results of transformation of one *species* into another by natural or artificial hybridization could be explained by the principles he had deduced from his own pea-hybrid experiments. Before commenting on this we should remind the reader again that, in Mendel's time, the word *species* did not mean what it means today. Although we have developed this idea already in our comments in section 2, it would be helpful to review these comments briefly here to refresh your memory. In Mendel's day the distinction between species and varieties was not as well worked out as it is now. The modern distinction between the two is based on the fact that hybridization between species is either not possible or results in a sterile hybrid. Unlike these hybrids between species, hybrids between varieties are fertile. There are many examples of plants classified in the past as separate species and since then reclassified as varieties of the same species. Kölreuter and Gärtner crossed many species (in the modern sense), but in general failed to obtain viable hybrids. The "species" that gave them fertile hybrids were merely varieties of the same species. For example, Mendel's peas are all now considered to belong to the same species (Berrie 1977, 76), though during Mendel's time they were not. Mendel raised some questions about the adequacy of the classification of peas in section 2 of his paper. Hence, what Mendel was discussing as the transformation of one species into another was really the transformation of one variety into another. In our discussion, we shall write the word *species* in quotes to indicate that it is not used in the modern sense.

What is meant by "transformation of one species into another?" In a general sense it means that a plant of "species" *B* becomes a plant of "species" *A* by the process of natural or artificial fertilization. Is this possible? The question was of utmost importance to nineteenth-century biologists who were interested in evolution. To answer the question they attempted to cross "species" *B* with "species" *A*, and then to cross the resulting hybrid again with "species" *A*. They repeated this operation many times, hoping that sooner or later they would recover "species" *A* from the progeny. We say that they "attempted" because many times they were unsuccessful in getting the hybrid, or if they got a hybrid it had low fertility. Even when successful, they never truly recovered "species" *A*, because some characteristics of "species" *B* always survived the process and were added to those of "species" *A*. Transformation in this sense was never successful. However, transformation can be accomplished if we restrict ourselves to the transfer of a limited number of characteristics. In that sense it is possible to transform one "species" into another, as Mendel showed and explained. In modern terms, this transformation process is called *backcrossing*. It is used by breeders to introduce a specific characteristic into an organism that is otherwise desirable—for example, disease resistance into crops such as wheat.

Gärtner carried out transformation experiments with many types of plants (lines 155–157) and found that this was the most difficult task he had encountered in hybridization (lines 147–148). He found out that the length of time it took to transform a "species" *A* into a "species" *B* depended directly upon how many fertilizations and generations it took to do this. It varied with the type of "species" and even with each individual plant (157–160). It varied also when he repeated the experiment (lines 160–161). In lines 161–170 he attempted to explain these variations by assuming that there was a force that carried characteristics from one plant to another, a force responsible for the "transformation of the maternal type." This force, said Gärtner, varies with each type of plant, and even with each individual plant. Hence, it is impossible to predict how much time is needed, that is, how many generations it will take to transform one "species" into another. Gärtner's suggestion that a force was involved in the phenomenon of transformation suggests to us that he believed in **vitalism.** Vitalism, a view prominent in the nineteenth century, assumed that the phenomena associated with living things are not the same as those associated with nonliving things and cannot be explained in the same fashion; that there are two separate sets of natural laws, one for living things and one for nonliving things. Vitalists always looked for vital forces to explain biological phenomema. Although Gärtner did this, Mendel did not.

In the rest of this section Mendel successfully explained Gärtner's results and set the phenomenon of transformation (or backcrossing) on a firm foundation. By using his idea that the hybrid produced "as many kinds of germinal cells as there are constant combinations made possible by the traits associated within the hybrid" (lines 174–175), he was able to explain the process of transformation without involving vital forces. Let us take the simplest example. Suppose the plants chosen for our transformation experiment differ in only one trait. Plant *A* is to be transformed into plant *B*. Plant *A* has the character *C*, plant *B* has the character *c*. We cross plant *A* with *B*. The hybrid that results from this cross has the breeding structure *Cc:*

Plant *A* (character *C*) crossed with Plant *B* (character *c*) = Hybrid of plants *A* and *B* (breeding structure *Cc*)

We now cross this hybrid with plant *B* to give us the first transformation:

Hybrid of plants *A* and *B* crossed with plant *B*

The hybrid will form two types of germinal and pollen cells, *C* and *c*. The plant *B*, on the other hand, will give only one kind of germinal and pollen cell, *c*. Hence, if we follow Mendel's symbolism, the offspring of the hybrid in this case will have as the breeding structure either *Cc* or *c*. The chances are very good (one-half) that we will recover a *c* type plant among the progeny of the cross, meaning that our transformation will be successful at the end of the first generation of backcrossing.

We now cross the hybrid with plant *A* in the second transformation:

Hybrid of plants *A* and *B* crossed with plant *A*

The hybrid will form the same types of germinal and pollen cells, *C* and *c*. The plant *A*, on the other hand, will give only one kind of germinal and pollen cell, *C*. The offspring of this cross will be *CC* and *Cc*, both of which will appear the same. In order to discover which of the offspring is true-breeding, the transformed one (*CC*), we have to raise a new generation of plants. Unlike in the first transformation, here we need two generations to be sure that the transformation is complete.

Suppose now that the plants chosen for transformation differ in two traits, *CD* and *cd* being their breeding structure. The hybrid resulting from the cross between plant *A* and plant *B* will have the breeding structure *CcDd* and will form four different germinal and pollen cells: *CD, Cd, cD,* and *cd*. When we cross the hybrid with plant *B,* we will have four types of offspring: *CcDd, Ccd, cDd,* and *cd*. Therefore, the chances that we will get a plant of type *cd* are one in four, which means that the chances that our transformation will be successful after the first generation of backcrossing are still very good indeed.

Mendel gave us (lines 182–204) the example of a transformation of two plants, *A* and *B,* that differ in three traits, *A, B,* and *C*. (Mendel unfortunately used the same symbols for the plant species and the traits by which these plants differed.) However, the same kind of reasoning used above permits us to calculate what the chances are that, after the first generation of backcrossing, we would get a plant of breeding structure *abc*. The chances are one in eight, and so there is, as Mendel said, "little likelihood that it [the transformed plant] would be missing among the experimental plants, even if only a fairly small number were raised, and transformation would thus be complete after two fertilizations" (lines 191–193). If by chance the transformed plant were missing, one would have to backcross one plant very similar to plant *B* and hope to get the transformation in the next generation of backcrossing.

It is clear, therefore, that the more traits there are to be transferred from one plant to another, the more generations; that is, the more time it will take to complete the transformation. This is "because the number of different germinal cells formed in the hybrid increases with the number of differing traits as a power of two" (lines 202–204). It also takes more time for transformation to occur if the number of plants among the offspring of each backcross is small (lines 192–200). So the observation of Gärtner that the time required to transform one "species" into another varies with the "species" is explained by Mendel. It takes more time for transformation to occur if both "species" are very different than if they are very similar.

In the next paragraph Mendel attempted to explain why transformation time varies with reciprocal crosses (*A* into *B* and *B* into *A*). In doing so, he showed that Gärtner was wrong when he criticized Kölreuter's idea that "the two natures in hybrids are in perfect equilibrium" (line 209). Instead, according to Mendel, this difference in transformation time had to do with the phenomenon of dominance. To understand this, let us assume that plant *A* has all the dominating characters and that plant *B* has all the recessive characters. We shall see that, in that case, it will take longer to transform *B* into *A* than it takes to transform *A* into *B*. The following will help you to understand the discussion:

> Plant *A* (all dominating characters, *CD*) crossed with Plant *B* (All recessive characters, *cd*) = Hybrids of plants *A* and *B* crossed with plants *B* and *A* (breeding structure *CcDd*)

Since both crosses will give us the same hybrid, both will produce the same four types of germinal and pollen cells: *CD, Cd, cD, cd*. The first backcross will yield different results depending on which transformation is made.

Let us now cross this hybrid with plant *B* to give us the first transformation:

> Germinal cells from hybrid crossed with pollen cells of Plant *B*
> (*CD, Cd, cD, cd*) *cd*

In the first transformation the germinal cells of the hybrid will unite with the pollen cells, type *cd* of plant *B*, yielding plants of types *CcDd, Ccd, cdD*, and *cd*. In this case, transformation of *A* into *B* would be terminated since the plant *cd* reappears after one generation of backcrossing.

Let us now examine the second transformation as shown below.

Germinal cells from hybrid crossed with pollen cells of Plant *A*
(*CD, Cd, cD, cd*) *CD*

In the second transformation the germinal cells of the hybrid will unite with the pollen cells, type *CD* of plant *A,* yielding plants of types *CD, CDd, CCD, CcDd.* All of these combinations will give the same appearance because of dominance. To know which plant among the offspring of this backcross is true-breeding, we have to self all the offspring. Though Mendel was well aware that he had to do this, he did not do it, possibly because he did not have the time (line 241–243). Hence, in this case, the transformation of *B* into *A* required one generation of backcrossing and one generation of selfing.

In our example, as well as the example given by Mendel, one variety had all the dominating characters and the other all the recessive. But this, as Mendel himself said (lines 244–245), is very infrequent. Most of the time, the dominating traits are distributed between the two plant varieties. But it is true that transformation takes a longer time when the pollen has the most dominating characters, because in this case we are not certain that the transformation is complete—the plant we selected as transformed might not be true-breeding.

No matter how long the transformation takes, that transformation is always possible. This led Gärtner to believe that plant species are stable within certain limits and therefore do not evolve in a haphazard way. Mendel concluded his paper by remarking that, though one should be careful in accepting Gärtner's opinion, there is no doubt that Gärtner's own experiments support such an idea.

Notes

1. It is believed from notes written by Mendel and discovered in 1965 that he may have come to the conclusion that species of willows could yield fully fertile hybrids, which in the next generation would segregate (Heimans 1970, 5:13–24).

2. We have underlined the word *elements* because it is only here, in his concluding remarks, that Mendel refers to cell "elements," to "antagonistic elements," and to "the material composition and arrangement of the elements that attained a viable union in the cell" (lines 91–92). People have attempted to interpret the word *element,* which Mendel used, as the equivalent of the word

gene used by the early geneticists. But this equivalence is tenuous, for Mendel never defined the nature of these elements. We also should note that Mendel says (lines 117 and 118) that "only those [elements] that differ separate from each other." This means that Mendel did not assume that elements which were the same separate. This assumption goes against one of the tenets of Mendelian genetics, the law of segregation. According to that law, each trait is associated with a pair of determiner factors, which, whether different from each other or not, segregate to individual reproductive cells as they are formed.

Epilogue

We have now traced Mendel's studies with hybrids from his original generative idea, that a law of hybrids must exist, through the discovery of that law, to his culminating creation of a quantitative theory of the formation of hybrids. It remains for us to explain how this theory may have come to be transformed by others, working at a later time, into the "Mendelian theory of heredity." We cannot offer here a formal, detailed history of this transformation because the information on which such a history must be based has not been assembled. Perhaps the most useful thing we can do at this time is to set down the outlines of a possible sequence of changes.

During the interval between the publication of Mendel's paper in 1866 and its recovery in 1900, certain discoveries in biology, in particular in the study of cells, were made. Biologists were able to make these discoveries because they overcame problems inherent in the study of cells: their small size and their opacity. Technical improvements in microscopes and in the preparation of tissues permitted scientists to search more effectively for the secrets of the cell.

Today we believe that any organism begins its life as a zygote; that is, a cell resulting from the union of two reproductive cells, the egg being the female contribution and the sperm being the male contribution. In this union the sperm penetrates the egg and its nucleus fertilizes the egg nucleus. But this explanation of how life begins is very recent. At the time Mendel wrote his paper, these ideas on reproduction were very speculative, to say the least. It is true the hypothesis that the bodies of animals and plants are composed solely of cells or cell products—the cell theory—dates from 1839, when T.A.H. Schwann and J. M. Schleiden proposed it. But biological ideas, like any other ideas, take time to be universally accepted; not until the last half of the nineteenth century was the cell theory established beyond reasonable doubt. It is also true that the presence of a nucleus, a spherical, rather dense body within the cell, was detected by Robert Brown in 1833. However, its function was not really known until 1885, when the story of cell reproduction was completed. It is also true that cell division was observed for the first time in 1835, but it was not recognized as a general phenomenon and was not understood for many years. We had to wait until 1873 to read the first reasonable published account of the complex nuclear changes, now called **mitosis,** that occur in body cells of all organisms. We had to wait another nine years before Walter Fleming showed conclusively that **chromosomes,** those intensely stained, usually rod-shaped bodies inside the nucleus, were not artifacts, but real; that their behavior at cell division was highly predictable; and that their number was constant in each cell of an organism.

There was one thing, however, that mitosis could not explain. This concerned the formation of the reproductive cells (gametes). If mitosis were involved in this process, the number of chromosomes would double in a zygote after fertilization. But, in fact it does not. The number of chromosomes per cell in the new generation is the same as that in the cells of the parents. So another process must be involved. There must be a mechanism for reducing the number of chromosomes at or before the formation of sex cells. In 1887 August Weismann proposed a hypothesis to account for the constancy of the number of chromosomes from generation to generation. During the second part of this type of cell division of both male and female gametes, there is a halving of the number of chromosomes. This is the process now called **meiosis.**

It is interesting to note that, though Mendel knew nothing about the internal structure of cells, his assumptions about pollen and egg cells and their role in the fertilization process were sound and were shown to be correct years later. Indeed, Mendel and other plant breeders before him played an important part, though an indirect one, in convincing the cytologists of the role of the nucleus in transmitting hereditary information. They repeatedly showed that in reciprocal crosses the maternal and paternal contributions were equivalent; that is, a female of variety *A,* fecundated by a male of variety *B* would generally give rise to the same manifestation of a particular characteristic in the progeny as a cross between a female of variety *B* and a male of variety *A*. Something present both in the egg and in the sperm had to be responsible for this fact. What could it be? In the egg cell the nucleus is surrounded by a large amount of fluid, called the cytoplasm, but the sperm is mostly a nucleus, with very little cytoplasm. Therefore, the hereditary information had to be in the nucleus.

Although biologists had proposed numerous theories of heredity before 1900, all were found to be wanting because they failed to explain many biological observations and because they did not link the particles of inheritance, which were assumed to exist, with any physical entity in the organism. The discovery of chromosomes and of their precise and regular behavior at cell division helped establish the hypothesis that the physical basis of inheritance is in the cell nucleus, specifically in the chromosomes. This hypothesis has stood the test of time. The chromosome theory of inheritance was the ultimate offspring of advances in cell studies during the last half of the nineteenth century. These important advances in cytology made in the last part of the nineteenth century permitted a physical understanding of the phenomenon of heredity.

As a result, by 1900 biologists were in a better position to attack the problems of development and heredity. It was time for two great lines of research to converge: the practice of breeding and the study of heredity. When Hugo de Vries, Eric Von Tschermak, and Carl Correns resurrected Mendel's paper, they saw it in a new perspective. They saw it as a study of heredity and not as a study of hybrids. As we said in the preface to this book, this new interpretation of Mendel was entirely possible because in the formation of hybrids, characters must be passed from parental plants to their offspring. This is basic to inheritance, and Mendel's very accurate data could therefore be used in a study of heredity. His paper, *reinterpreted* in this light by others, became the founding document of the science of genetics.

However, this new science, which developed out of the fusion of these two lines of research and which we call Mendelian genetics, proved to have limitations, just as Mendel's work on hybrids had. Mendelian genetics, as opposed to Mendel's work, answered the question, What is it in the sperm and egg that is responsible for the transmission and development of characters? by inventing the concept of the gene. As research continued, the locations of individual genes on the chromosomes were determined by breeding experiments, and maps of such locations were constructed. Although such research contributed much more to our knowledge of heredity, it, too, gradually came to its limits.

To get beyond this limit it proved necessary to do once again what

Mendel had done: look for ways to draw upon the concepts and methods of other physical sciences and find ways to apply them to biological problems. In Mendel's time it was possible for one person to build the bridge between these two areas, but this was no longer possible. The relevant sciences, genetics, biochemistry, and biophysics, had each become too complex and too independent of each other for any person to tie them together. As a result, the new answers to the old sperm and egg question were given by teams of two or more individuals, each drawn from one of the disciplines. Out of these fruitful collaborations has come the whole new science of molecular biology with our knowledge of the structure of DNA, the basic molecule of inheritance, and the way it and other molecules function in all living things.

If the pattern of the past holds true in the as yet unguessed and unanticipated future, in time these studies will also reach their limits. Then, once again, other men and women will search for ways to bridge across still unknown areas of knowledge to carry on the endless search which generates new questions with each new set of answers.

Appendix A

Fertilization in Plants

In the chapter "Botany of Peas," we described the process of pollination, but said little about the process of fertilization in plants. The reason why we did this is that it is not necessary to know in detail how a plant is fertilized in order to understand what Mendel was doing in his experiments. In his time little was known about it. Even the fundamental question of how many pollen grains are needed to fertilize one egg cell had not been answered. Mendel assumed—and he was right—that only one was needed. In fact, the process of fertilization in plants is very complex, more complex than in animals. It is described in this appendix.

Just as in other organisms a pea plant starts as a *zygote,* a fertilized egg. The zygote, the first embryonic cell of the new plant, results from the union of a sperm nucleus with an egg nucleus within the egg cell. Where do these two nuclei come from? What is the importance of this union? To answer these questions we have to look at the process of sexual reproduction in plants. This can be divided into three distinct steps:

1. Formation of pollen grains and formation of egg cells
2. Pollination, the transfer of pollen grains from anther to stigma
3. Fertilization, the union of sperm and egg nuclei

Formation of Pollen Grains

If you cut a young anther crosswise (fig. A.1), you can easily see within it several chambers, called *pollen sacs.* Using a microscope, you can readily see that each pollen sac contains clusters of cells with large nuclei. These are known as *microspore mother cells* (fig. A.2, step 2). Like the other cells of the plant they contain two sets of chromosomes ($2n$), one paternal set and one maternal set. These cells are called *mother cells* because, through two cell divisions, they give rise to four daughter cells, called *microspores* (step 3). During this division process, called *meiosis,* the

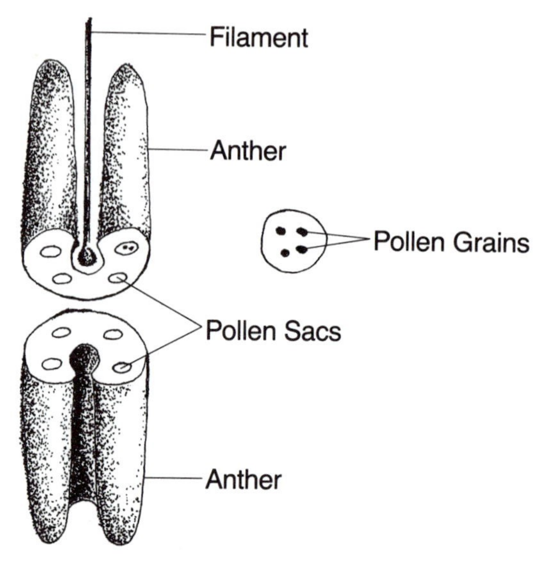

Fig. A.1. Diagram of a Sectioned Young Anther

chromosome pairs separate, each microspore receiving only one chromosome of each pair. Each microspore then develops into a pollen grain (step 4) containing half the number of chromosomes present in the microspore mother cells ($\frac{1}{2} \times 2n = n$). The nucleus of each pollen grain then divides, forming two nuclei (step 5), which will have very different and very specific functions in fertilization. One is the *tube nucleus,* the other is the *generative nucleus.* When these nuclei have been formed, the anther bursts open and the pollen is shed. Having outlined the formation of the male reproductive cells, we turn next to the formation of the female reproductive cells. When we have completed this, we will discuss their union in fertilization.

Formation of the Egg Cells

The eggs originate in the ovary from special cells, the future ovules. These first appear as tiny knobs on the inside of the ovary wall (fig. A.3, step 1). As an ovule develops it is raised from the ovary wall by a short stalk. It is through its stalk that the ovule, the future seed, receives its nourishment. Inside each ovule there is a special cell, the megaspore mother cell (step 2), which has two sets of chromosomes ($2n$). As the ovule develops this cell divides *meiotically,* the same way as the microspore mother cell of the pollen. This results in a set of four megaspores (step 3), each having half

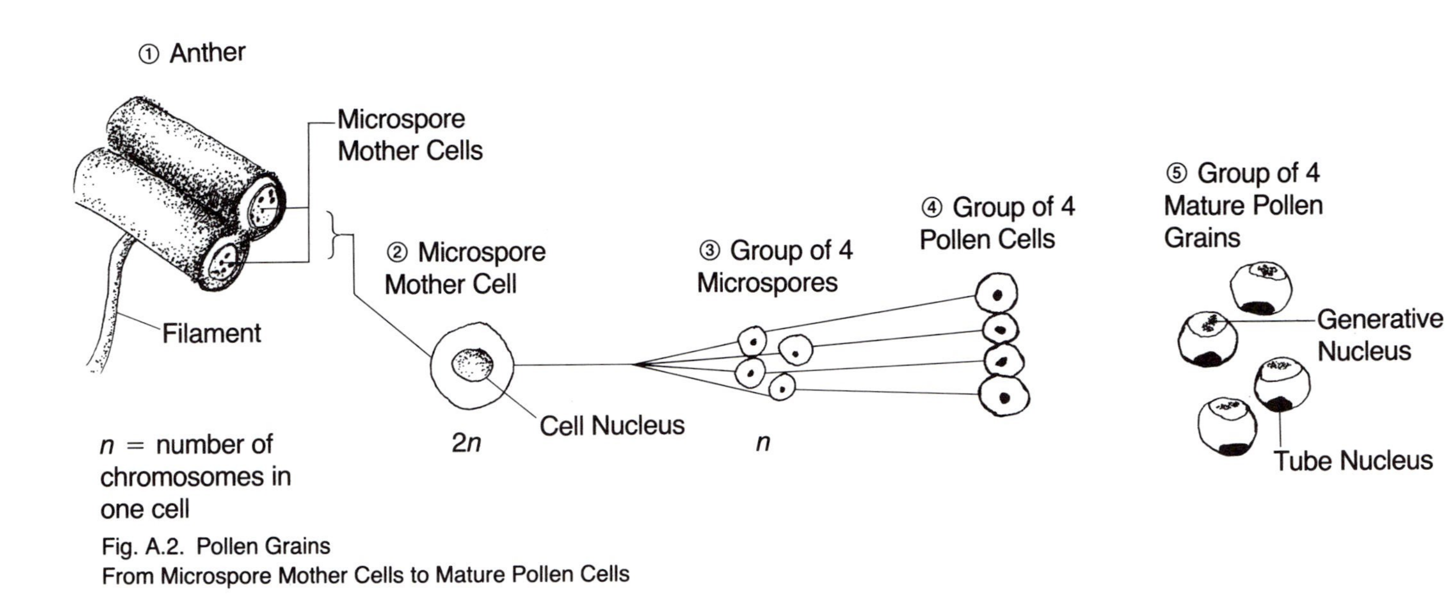

Fig. A.2. Pollen Grains
From Microspore Mother Cells to Mature Pollen Cells

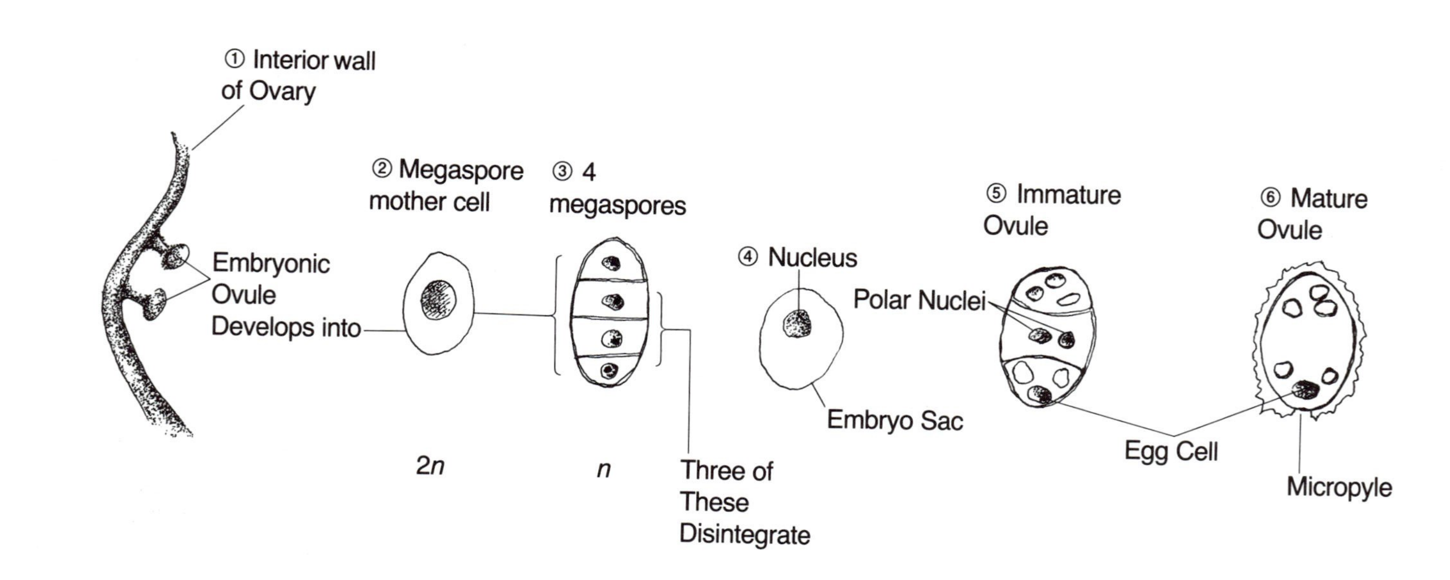

n = number of chromosomes in one sex cell

Fig. A.3. Egg Cells
From Megaspore Mother Cells to Ovules

the number of chromosomes (n) as the megaspore mother cell had. Of the four megaspores produced, three disintegrate. The remaining megaspore enlarges rapidly and forms a sac called the *embryo sac* (step 4). The megaspore nucleus then undergoes three successive divisions resulting in 8 nuclei within the embryo sac (step 5). Only three of these nuclei will be involved in the process of fertilization. The two nuclei in the center of the embryo sac are called *polar nuclei*. The one close to the micropyle is the *egg nucleus*. Two protective layers form around the ovule and enclose it completely except for a tiny pore, the *micropyle*, on the lower side of the ovule (step 6). It is through the micropyle that the sperm nucleus from the pollen will enter the ovule and unite with the egg nucleus when fertilization takes place.

The stage is now set for the conclusion of the drama, which takes place after pollination has landed a pollen grain on the stigma of the plant: germination of the pollen grain and fertilization. For fertilization to be successful the pollen grains and the ovules must both be mature at the same time. This is what happens in peas and makes it possible for self-fertilization to occur with the flowers unopened.

Germination of Pollen and Fertilization

Once a pollen grain has landed on the stigma, it immediately germinates (fig. A.4, step 2) and a pollen tube begins to grow out of it and down through the inside of the style toward the ovary. During the growth of the pollen tube the generative nucleus divides, producing two sperm nuclei, which move down with tube (step 3). Eventually, the tip of the pollen tube reaches the ovary and, passing through the micropyle (step 4), enters the mature ovule. At this time, one of the sperm nuclei fuses with the egg nucleus to form the zygote, the first cell of the new plant. The other sperm nucleus fuses with the two polar nuclei (not shown in the diagram) to begin the formation of an endosperm cell. These two processes always take place simultaneously in a double fertilization.

The fusion of the sperm nucleus with its (n) chromosomes and the egg nucleus with its (n) chromosomes restores the original number of chromosomes ($2n$) characteristic of that plant. The nucleus of the endosperm cell ($3n$) immediately undergoes a series of divisions that result in the development of the endosperm, an available food source for the growing embryo. The endosperm itself may be entirely consumed by the developing embryo, as in peas, or it may persist in the fully ripened ovule, as in all the grasses, where it will be used by the embryo during the germination process. The development of the seed was covered in the chapter "Botany of Peas."

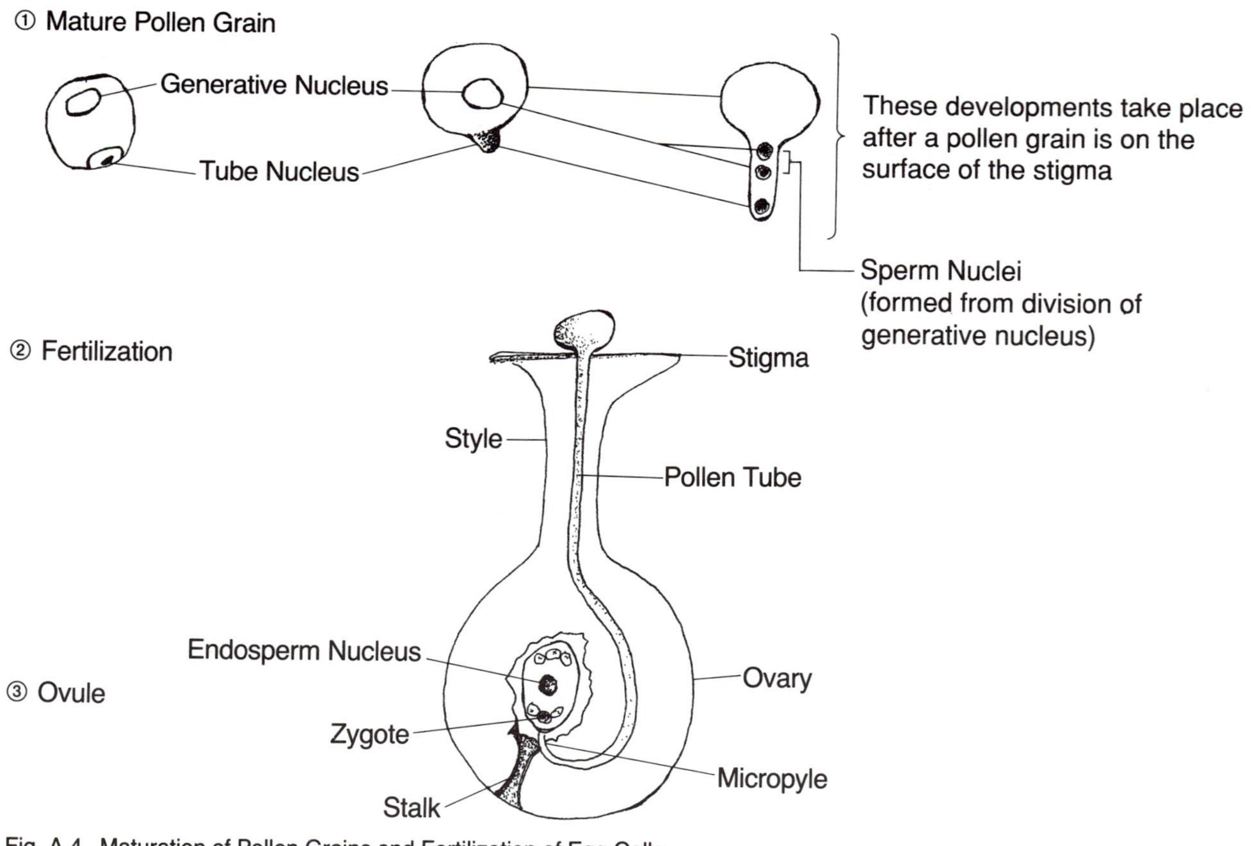

Fig. A.4. Maturation of Pollen Grains and Fertilization of Egg Cells

The Biological Importance of Sexual Reproduction

The main feature of sexual reproduction in plants as well as in animals is the production of offspring, each of which is unique, hereditarily different from its parents, and different from any other except in the case of identical twins. The origin of this diversity lies in two processes: meiosis, by which the reproductive cells, the gametes, are produced; and fertilization, which is also a random process.

Each gamete from an individual organism is different from all other gametes of that organism. The reason for this is that, when a gamete is formed by meiosis, it receives a unique set of chromosomes consisting of one chromosome from each pair carried by the parent organism. It is a matter of chance which chromosome of a pair is received by a newly formed gamete. Because of the large number of chromosomes in most kinds of plants, the chance of two gametes from the same individual receiving exactly the same chromosomes is very small. Since each zygote results from the chance union of a unique sperm cell and a unique egg cell, it will be hereditarily different from either of its parents and from any other zygote that issues from those parents. Meiosis is the fundamental reason why there is such diversity among the offspring of any couple of organisms. Because of the importance of meiosis in producing diversity among organisms, we believe that an understanding of this process is fundamental. Therefore, we have developed the concept in detail in appendix B. Again we must emphasize that Mendel did not know about meiosis, nor about the details of the process of fertilization that we have discussed above. It is interesting to speculate about what changes he might have made in his experiments and/or his explanation of his results had he had this knowledge.

Appendix B

Meiosis

Meiosis is basically the same in all sexually reproducing organisms and offers an elegant illustration of one of the fundamental assumptions underlying Mendel's work, as stated in his concluding remarks: "One might assume, however, that no basic difference could exist in important matters since *unity* in the plan of development of organic life is beyond doubt" (section 11, lines 141–143)

To help you understand the process of meiosis and some of its benefits for sexually reproducing organisms, we shall use some portions of a standard deck of playing cards to create a model of the process. In this model, each card represents a chromosome. Since a pea plant has seven pairs of chromosomes, we shall use only seven cards from each of the four suits in the deck. We have chosen to use the ace, king, queen, jack, ten, nine, and eight from each suit. The red cards represent the chromosome makeup of the male parent of a child, each pair of red cards representing a pair of homologous chromosomes. The black cards represent the chromosome makeup of the female parent of the child, each pair of black cards representing a pair of homologous chromosomes.

Just as each of us has two parents and four grandparents, each plant has two parents, a pollen plant and a seed plant, and each of these has two parents, the two paternal pollen and seed grandparents and two maternal pollen and seed grandparents. The four suits represent the chromosomal contribution of the grandparents to the parents of the child and, as we shall see, their specific chromosomal contribution to their grandchild. Thus, the spades represent the chromosomal contribution of the maternal seed grandparent; the clubs represent that of the maternal pollen grandparent; the diamonds represent the chromosomes contributed by the paternal seed grandparent; and the hearts represent the chromosomes contributed by the paternal pollen grandparent. We have expressed these relationships in table B.1. The seven pairs of chromosomes making up the complement of the seed parent consist of a matched set of seven spades and seven clubs.

Table B.1. Sorting of Chromosomes in Meiosis and Fertilization

Meiosis

Chromosome Complement of the Parental Seed Plant (2*n*)		Chromosome Complement of the Parental Pollen Plant (2*n*)	
From grandparent of the seed plant ♠	From grandparent of the pollen plant ♣	From grandparent of the seed plant ♦	From grandparent of the pollen plant ♥
Ace	Ace	Ace	**Ace**
King	**King**	King	**King**
Queen	**Queen**	**Queen**	Queen
Jack	Jack	**Jack**	Jack
10	**10**	**10**	10
9	**9**	9	**9**
8	8	**8**	8

Egg	Sperm
Ace ♠	Ace ♥
King ♣	King ♥
Queen ♣	Queen ♦
Jack ♠	Jack ♦
10 ♣	10 ♦
9 ♣	9 ♥
8 ♠	8 ♦

Fertilization

The formation of one specific zygote with 14 chromosomes in 7 pairs (2*n*)

Ace ♠	Ace ♥
King ♣	King ♥
Queen ♣	Queen ♦
Jack ♠	Jack ♦
10 ♣	10 ♦
9 ♣	9 [illegible]
8 ♠	8 ♦

NOTE: The card-chromosomes in boldface are the chromosomes of each pair passed by chance to the gamete.

In the same fashion the seven pairs of chromosomes of the complement of the pollen parent consist of a matched set of seven diamonds and seven hearts. You will notice that the symbols for some of the card-chromosomes are in boldface type. The significance of these will be explained below.

In our example, the zygote had received by chance three spades and four clubs on the maternal side and four diamonds and three hearts on the paternal side. The combination of card-chromosomes from the "mother" could equally have been five spades and two clubs or any other combination that added up to seven cards. The same would be true of the combination of card-chromosomes from the "father's" side. Remembering that these cards of the different suits represent the contributions of the grandparents to this zygote, we can see how meiosis is one of the main sources of the great diversity that can exist among offspring of one couple. We will explore the nature of this diversity in the following paragraphs.

When an egg is formed it can contain only one of each pair of the total chromosomes that the mother received from her parents. For each pair of chromosomes, the chances that one chromosome will appear in an egg are the same as for the other. We have illustrated this in our card model by setting in boldface type one chromosome-card of each pair at random. The result is a unique set of chromosomes that will be present in one egg. We can attach a certain probability to each set in the following manner.

The probability is one-half (0.5) that the ace-of-spades chromosome will be the one included in the egg. The probability that it will be the ace-of-clubs chromosome is also one-half. The probability of either king (spades or clubs) being present in the egg is also one-half, and the same holds true for each pair. Since each choice is independent of the others, the chances that "our egg" will contain both the ace-of-spades chromosome and the king-of-clubs chromosome is the product of their independent probabilities of being present. This is $\frac{1}{2} \times \frac{1}{2} = \frac{1}{4}$ or $(\frac{1}{2})^2$. And the probability that the arrangement could be just the reverse is also $\frac{1}{4}$ (0.25).

Since there are seven chromosome-cards in each set, the probability of getting any one particular set of card-chromosomes in the egg is $(\frac{1}{2})^7$ or $\frac{1}{128}$. This is true because there are 128 different possible combinations, each of which is equally probable. In the case of our example, there are 128 different ways in which we could select gametes from an individual without repeating ourselves.

We can pursue the same argument with the formation of an individual sperm and come to the same conclusion. There would be 128 different possible combinations of card-chromosomes, each of which would have the same probability of appearing in a particular card-chromosome sperm.

Since fertilization, the meeting of a sperm and an egg, is also a random or chance process, we can calculate how many different combinations are possible without duplication simply by multiplying 128 × 128. The result is 16,384. Hence, the chances of any one egg meeting any one sperm in our card-chromosome analogy is one in 16,384. Thus, with only 14 chromosomes in 7 pairs there are 16,384 different combinations that could occur by chance. Each of these combinations in the zygote would result in an individual with a unique combination of characteristics. This is an indication of the way in which sexual reproduction contributes to the diversity of the biological world.

Since we are always interested in seeing how ideas such as these apply to us, let us make a calculation for ourselves. We have 46 chromosomes in 23 pairs in our body cells and 23 chromosomes in our sex cells. Hence, using the same type of reasoning as we used in the above example, we can deduce that the chances that a child will have the same chromosome complement as its sibling are $1/2^{23} \times 1/2^{23} = 1/2^{46}$ or $1/8{,}388{,}608 \times 1/8{,}388{,}608$ or 0.00000000000007. Even this very small number is too large because chromosomes exchange parts during the formation of sex cells, which increases even more the diversity between individuals. The chances that two children of the same parents have the same hereditary makeup are nil. If this is true for children of the same parents, the chances are even smaller for children of different parents.

Appendix C

Schedule of Mendel's Monohybrid Experiments

Mendel never gave a time schedule of his experiments. Hence, it is hard to keep track of the years in which the experiments were carried out. To help the reader we have made one for his monohybrid crosses.

Year 1, 1856

Spring: Mendel crosses his parental varieties
Late Summer: Production of seeds containing hybrid embryos
Fall: Harvest of seeds
Winter: Mendel collects data for seed traits. (hybrid generation). He cannot collect data for adult traits at that time; to obtain these data he has to plant seeds the and wait until the plants mature.

Year 2, 1857

Spring: Mendel plants hybrid seeds
Early and late summer: Production of seeds of the hybrid plants. Mendel collects data for adult traits of the hybrid plants (hybrid generation):

Exp. 3. All plants have grey-brown seed coat
Exp. 4. All plants have smooth arched pod
Exp. 5. All plants have green unripe pods
Exp. 6. All plants have axial flowers
Exp. 7. All plants are tall

Fall: Mendel harvests seeds
Winter: Mendel collects data for seed traits and calculates ratios (first generation from hybrids):

Exp. 1. 3 round to 1 angular
Exp. 2. 3 yellow to 1 green

Year 3, 1858

Spring: Mendel plants seeds of the hybrids
Early and late summer: Mendel collects data for adult traits (first generation from hybrids):

Exp. 3. 3 grey brown to 1 white seed coat
Exp. 4. 3 smooth to 1 constricted arched pod
Exp. 5. 3 green to 1 yellow unripe pod
Exp. 6. 3 axillary to 1 terminal flower
Exp. 7. 3 tall to 1 dwarf

Fall: Harvest of seeds
Winter: Mendel collects data for seed traits and calculates ratios (second generation from hybrids). For each seed trait, Mendel obtains the following ratio:

1 true-breeding dominant
2 hybrid dominant
1 true-breeding recessive

Year 4, 1859

Spring: Mendel plants seeds from the first generation from the hybrids (hybrid plants having naturally self-fertilized)
Early and late summer: Mendel collects data for adult traits (second generation from hybrids). For each of those he obtains the same ratio he got for the seed traits and calculates ratios:

1 true-breeding dominant
2 hybrid dominant
1 true-breeding recessive

Appendix D

Problems with Seed Coat

In his trihybrid experiment Mendel studied what happened when the parental plants differed in three traits: seed shape, cotyledon color, and seed coat color. Mendel said that of all his experiments this one required the most time and effort, but he did not tell us why. He gave us the results of a crucial experiment and his interpretation of these results, but very little explanation of how he obtained them. In this appendix we describe in some detail the difficulties Mendel had to overcome in carrying out this important experiment, which required tremendous planning, accurate record keeping, and considerable mathematical and deductive ability.

Some of the reasons why this experiment was so difficult are obvious, others are not. For example, it is obvious that, in experiments involving three traits, a very large number of plants or seeds have to be examined, classified, recorded, and counted. However, by itself this should not have made the experiment as difficult as Mendel seems to suggest. The difficulty, as we have said in the text, lay in his choice of three seed traits, a choice that seemed at first glance to be a good one but in fact was not. Mendel could easily determine each of those seed characteristics at the end of the growing season by looking at each individual seed. Two of the seed traits, color of the cotyledons and seed shape, were the traits he had used in his dihybrid experiment and he knew how they behaved. However, in his trihybrid experiment, he included another seed trait, seed coat color. This brought serious difficulties—to which Mendel never referred.

The first difficulty is that the color of the cotyledons, which can easily be seen through a transparent (white) seed coat, cannot be seen through a grey-brown seed coat, which is opaque. Unfortunately, because of the way Mendel selected the parental plants of his trihybrid, most of the seeds in his trihybrid experiment were opaque. How, then, did he determine the color of the cotyledons? Did he soak the seeds to detach the seed coat from the embryo? Did he cut through the seed coat so that he could see the cotyledons? Mendel did not tell us. However, he must have found a way to overcome this difficulty. One thing is certain: whatever his method was, the determination of the color of the cotyledons for each seed must have required a considerable amount of care and effort.

The second difficulty, which we have already mentioned, arises because the pea seed has two distinct parts, the seed coat and the embryo. The seed coat is part of the parent (seed) plant, while the embryo with its cotyledons (the first two leaves) is part of the next generation. This presents a record-keeping problem for a plant breeder such as Mendel. Examining a seed carefully, he could determine what the seed shape and the color of the cotyledons were. Both are characteristics of the embryonic plant. He could also determine the color of the seed coat of that particular seed, but seed coat was not part of the embryo enclosed within it. When Mendel looked at the seed coat he was looking at material that was part of the maternal generation. In order to associate the three seed traits for each generation, he had to keep a record of two of the three seed traits for each individual plant at the beginning of the season and of the third at the end of the season.

In planning this trihybrid experiment Mendel brought to bear all of the experience he had accumulated in planning, carrying out, and interpreting his previous breeding experiments. However, in this experiment the components of his previous experiences were merged, extended, and refined in planning the sequence of individual subexperiments and in combining the interpretations of their individual results. What Mendel did here was far beyond the capacity of a merely competent plant breeder. It required an unparalleled grasp of the essentials of the breeding process and an unrivaled ability to extract meaning from the data.

In the following, somewhat compressed, notation we have endeavored to describe the procedure by which Mendel carried out this complex experiment and kept track of his data.

Time	Procedure
1st year	Predicts results of a cross of two plants, seed plant *ABC* × pollen plant *abc* = *AaBbCc*
Spring 1859	Plants the seeds of seed plant and pollen plant
Summer 1859	Makes the cross
Fall 1859	Harvests seeds
	Results: 24 round seeds with yellow cotyledons and brown seed coat (brown seed coat is part of maternal tissue from seed plant)
	Records data in notebook and on a bag in which hybrid seeds are stored

2d year

Spring 1860	Plants 24 hybrid seeds separately in a special bed in the monastery garden
Summer 1860	Grows 24 hybrid plants and allows all flowers of each plant to self-fertilize
Fall 1860	Harvests seeds, which Mendel calls the first generation from the hybrid Results: 687 seeds, some of which are round, some wrinkled. Also, some have yellow cotyledons and some have green cotyledons. *All* have brown seed coats (brown is dominant over white)

3d year

Spring 1861	Plants each of the 687 seeds separately in a special garden bed, recording in a notebook the planting location of each seed together with its shape, cotyledon color, and seed coat color
Summer 1861	Grows plants from seeds (not all produced plants; only 639 plants grew to maturity) and allows all flowers of each plant to self-fertilize
Fall 1861	Harvests seeds from each plant into separate, labeled bags. Mendel calls these seeds the second generation from the hybrid Records (1) whether each plant produced only round seeds, only wrinkled seeds, or a mixture of both; (2) whether each plant produced only yellow seeds, only green seeds, or a mixture of both colors; and (3) whether each seed has a brown seed coat or a white one Note A: These data enabled Mendel to tell what the breeding structure of each of the 639 plants was with regard to the first two characteristics. (1) If a plant produced only round seeds, it must have been true-breeding for round shape and hence had the breeding structure *A* for this character. In the same way, if a plant produced only wrinkled seeds, it must have had breeding structure *a* for this character. And if the plant produced both round and wrinkled seeds, it must have had the breeding structure *Aa*. (2) In the same way the breeding structure for cotyledon color could be deduced as *B, b,* or *Bb*. Note B: In order to determine the breeding structure of

each of the 639 seeds for seed coat color, it was necessary to carry the experiment through another generation, which Mendel did the following year.

4th year

Spring 1862	Sows a sample (probably 10 seeds from each of the 639 bags of seeds), recording locations of each of the 6,390 seeds
Summer 1862	Grows plants and allows all flowers of each plant to self-fertilize
Fall 1862	Harvests separately the seeds from each plant and checks to see whether all seeds have brown coats or all have white coats or are a mixture of both Note A: The interpretation of these data parallels that for the other two traits—if all seeds were brown the breeding structure of the plant and the embryo in the seed from which it came from had to be *C* for seed color; if all seeds were white, the plant and the embryo in the seed from which it grew had the breeding structure *c;* if the seeds were of both colors, the breeding structure of the plant and the embryo in the seed from which it came were *Cc*. Note B: Since Mendel already knew from the data of the third year what the breeding structure of each seed was for shape and color traits, he now could establish the breeding structure of each of the 639 seeds for all three traits.

These data, summarized in his report of the experiment, enabled Mendel to establish that the trihibrid formed progeny in conformity with the law he had discovered in his monohydrid and dihybrid crosses.

A Final note: In the text of the trihybrid experiment, just before the summary of his data, Mendel wrote: "Of the plants grown from them [the 687 seeds from the hybrids], 639 bore fruit in the following year and as further investigation showed they comprised. . . ." That short expression "as further investigation showed" is what it has taken all of the pages of this appendix to explain. Mendel is not alone in making statements that pass over sections of an experiment in this way. Indeed, they are common in scientific papers written to communicate ideas to other scientists in a particular field. However, even there what is left out may be critical to understanding some central idea—which may not be as obvious to the reader as it was to the author. When such a paper is to be interpreted by scientists in another field or, worse still, by a nonscientist, these gaps create problems such as the one we have dealt with here.

Appendix E

Where Is the Bias in Mendel's Experiments?

One of the great controversies surrounding the life and work of Mendel has centered on whether or not he falsified some of his data. In 1936 the eminent biologist and statistician R. A. Fisher proposed, on the basis of a statistical analysis of Mendel's data, that either Mendel (or more likely one of his assistants) had "cooked" his data, which seemingly were too close to the expected values. This was heresy! The idea that Mendel—a priest, an Augustinian monk, and later an abbot of a monastery—could do such a thing was repugnant. The idea that one of the great biological scientists of the nineteenth century could do so was abhorrent. Such behavior ran contrary to the ideal of total honesty in science and in religion. Therefore, it must have been an assistant and not the great man.

Nonetheless, some faint aura of suspicion still clung to Mendel. After all, Fisher was a master statistician, an international authority in the field of experimental design and the analysis of data. For many, if Fisher said the data were too good to be true, then they were—and that was that. There the matter rested for some years, until finally some students of Mendel's work began to reexamine the data to see whether they really were too good or whether there might be something wrong with Fisher's calculations.

As a result of this reexamination, in which the two authors of this book played minor parts, Mendel's reputation was indeed restored. The same statistic responsible for reputation lost was also responsible for reputation regained. That statistic is known as chi-square. In this appendix we introduce you to this useful tool in a way that smooths out many of the difficulties you might encounter in an introductory text. When you understand what the statistic is and how it is used, you will appreciate better the controversy that surrounded Mendel and how that controversy was resolved.

In section 6, where Mendel reported the results of his monohybrid crosses, he found that among the offspring of the hybrids showing the dominating trait there was one offspring that bred true for each two that did not. His experiments with all seven traits gave approximately the same ratio, except for experiment 5, which he repeated. Mendel said: "Experiment 5, which showed the greatest deviation [from the 2 to 1 ratio], was repeated, and instead of 60–40 [the ratio is 1.5 to 1], the ratio of 65:35 [or 1.86 to 1] was obtained. *Accordingly, the average ratio of 2:1* [2 hybrids to one true-breeding] *seems ensured*" (lines 42–44). This last sentence of Mendel's is very interesting since one could ask what would have happened if on his second trial of experiment 5 he had found a ratio further from 2 to 1 than the one he found on the first trial. Would he have repeated that experiment once more, and once more again, until the ratio came out "right?"

Probably not, because Mendel knew that in this kind of biological experiment, even those involving hundreds of individual plants, he could not expect to obtain results that were exactly correct. His data might appear to be getting closer and closer to some perfect ratio as the number of individuals got larger and larger, but still would not reach the exact ratio. If every seed in every experiment produced a plant that blossomed and produced exactly the same amount of seed, then seed-trait experiments would yield perfect data. But this is never the case. There are always some seeds that do not germinate, and there is no way of knowing beforehand which and how many will fail. Some of the plants that are produced fall victim to drought or to disease or to damage by insects or animals, and again there is no way of knowing how many will fail in these ways. Even if the plants produce flowers there may be variations in the completeness of fertilization or in the development of seed pods or the ripening of the seeds. It is chance factors such as these (and there are still others) that result in data that are less than theoretically perfect. As the numbers of individuals are increased and the results averaged, the influences of individual failures due to chance events on the final results are smoothed out but not totally eliminated. This always leaves the experimenter with the problem of deciding when his or her results are good enough to accept as evidence of the correctness of his or her ideas.

In Mendel's experiment in section 5, the question was whether, in terms of his hypothesis, deviation of the observed results from the predicted ones was within limits set by chance alone. In Mendel's time the science of probability had been born, but was not as sophisticated as it is now. No statistical tool had yet been invented to help Mendel make this type of decision.

Today we have one. It is called the chi-square test and it helps us to determine whether an observed ratio is an acceptable example of an expected ratio. In other words, we use chi-square to determine "goodness of

Table E.1. Chi-square Test for Experiment 5 of the Second Generation of Hybrids

Class	Observed	Expected	Deviation		
Plants with pods			$(O - E)$	$(O - E)^2$	$(O - E)^2/E$
a. Green and yellow	60	66.7	−6.7	44.89	0.672
b. Green	40	33.3	+6.7	44.89	1.348
Sum = Σ =	100	100	0.0	89.78	2.020

fit" of observed values obtained in an experiment to the predicted or theoretical values.

Let us think for a bit about this particular tool, chi-square. What sort of characteristics should it have? First of all, it ought to be easy to understand why it is calculated as it is and it ought to be easy to calculate; that is, the mathematics of using it ought to be simple and straightforward. Also, the way it behaves ought to seem logical and reasonable. Finally, it ought to be really useful in helping us to make decisions.

We want the statistic (the number we finally obtain) to become smaller the smaller the difference between what we observe in our data and what we expect to get ideally, and vice-versa. This suggests that we start by taking the difference between the two for each of the observations we make. Let us take as an example the data from Mendel's own experiment 5 that we mentioned earlier and see how it works.

You remember that the first time he carried out the experiment he obtained among the offspring of the hybrid 40 plants with green pods only and 60 plants with both green and yellow pods. What were the expected values? Since there were 100 plants and since the expected ratio is 1 to 2, the expected number of plants with green pods only is ⅓ × 100 or 33.3 and the number of plants with green and yellow pods is ⅔ × 100 or 66.7. Now let us set these and his actually observed values down in a small table (table E.1) and then subtract the expected values from the observed values and see what we get. If we add these differences $(O - E)$ we get zero, which does not give us the information we want, yet we can see that there are real differences.

Now, if these + and − signs were just not there, this might help. There is an old mathematical trick to take care of this, − square both differences. We then have for class *a* 44.89 and for class *b* 44.89, and the sum is now 89.78. This is better, but there is still a problem. Neither the +6.7, nor the 44.89 we converted it into has the same importance with respect to the expected value of 66.7 as does the −6.7 or 44.89 to the other expected value of 33.3. We can adjust for this difference in relative values if in each case we divide $(O - E)^2$ by the corresponding value of E, the value we expected. Thus, for class *a* $(O - E)^2/E$ = 44.89/66.7, which equals 0.672, and for class *b* $(O - E)^2/E$ = 44.89/33.3, which gives 1.348. Now, if we add these two values 0.672 and 1.348 together we get 2.020. This last number is our statistic, chi-square. If we write out what we have done in shorthand symbolic form, this is what we get

$$\chi^2 = \Sigma \frac{(O - E)^2}{E}$$

Table E.2. Chi-square Values and Associated Probabilities for the First Three Degrees of Freedom

d.f. \ p	.99	.95	.90	.80	.70	.50	.30	.20	.10	.05	.01
1	0.000	0.004	0.016	0.064	0.148	0.455	1.074	1.642	2.706	3.841	6.635
2	0.020	0.103	0.211	0.446	0.713	1.386	2.408	3.219	4.605	5.991	9.210
3	0.115	0.352	0.584	1.005	1.424	2.366	3.665	4.642	6.251	7.815	11.345

Impressive, but not mysterious because you have followed each step of reasoning in its development.

Now that we have a statistic that behaves the way we want, how can we use it to decide how Mendel's data affect his hypothesis? What could it say? One thing it might say would be that these data are so far away from predicted values that the 2 to 1 ratio hypothesis ought to be rejected and something else looked for. That kind of judgment would arise if the $(O - E)$ differences got too large. But how large is too large? On the other hand, the statistic might say these data are so close to the predicted values that something is fishy. Maybe the data were either deliberately or unconsciously modified to make them fit the theoretical values very closely. That kind of judgment would arise if the $(O - E)$ differences were too small. But again, the question is how small is too small?

From this it is evident that there must be a range of values of $(O - E)$ where there is no reason to reject the 2 to 1 hypothesis but also no reason to question the data as perhaps biased. In that wide middle ground the variations in the data are such as might arise from those chance factors we considered at the outset. But there is one set of factors that we have to consider before we can use our statistic to make the kinds of judgments for which it was designed. Those are errors we can relate to problems in making accurate counts of the peas. We can illustrate this by using the following example.

Suppose that a friend of yours has put ten peas in a jar and asks you to withdraw them in two attempts. On the first attempt, you will have complete freedom to withdraw anywhere from one to nine peas. But once you have determined the first group, the second one is also determined. In other words, if you took four peas from the jar on the first attempt, you are bound to take six on the second attempt. Here we have only one degree of freedom of choice.

The same is true in Mendel's experiment relating to pod color, where he had only one degree of freedom because there are only two classes: pods with yellow and green peas and pods with only green peas. Here the number of degrees of freedom is always one less than the number of classes. We need this idea to interpret an obtained value of chi-square, because the way those values are distributed in table E.2 below varies with the number of degrees of freedom present in the data. The values in the table also vary with the probability that our chi-square would differ from a theoretical value to a certain extent due simply to the influence of chance factors such as those mentioned earlier.

Having decided that there is only one degree of freedom, we select the top horizontal row of figures, which represents the values of chi-square for one degree of freedom. We then match the chi-square value that we have obtained, 2.020, with those in the table and we find that it falls between 1.642 and 2.706. (table E.2). This indicates a probability of between 10 and 20 percent, which means that the deviation between the observed and the expected results would be this great or greater in about 10 percent of the crosses of this nature, provided only chance was operating. Had the chi-square value been 2.706 or greater, the chances of getting such large deviations between the observed and the expected, the hypothesis being correct, are 10 percent or less. Had the chi-square value been 3.841 or greater, the chances of getting such large deviations between the observed and the expected, the hypothesis being correct, are 5 percent or less. The 5 percent point in the table is usually chosen as an arbitrary standard for determining the significance of goodness of fit. It is a reasonable standard, for at this point there is one chance in twenty that a hypothesis that is true will be rejected, and there is one chance in twenty that a hypothesis that is false will be accepted as true. Thus, the chi-square test gives us an estimate of the chance of error in our decision to reject or not to reject a hypothesis. However, it is up to the researcher to accept or not accept the results of an experiment with respect to the hypothesis.

Let us go back to Mendel's experiment. When we entered the table of chi-square with one degree of freedom, we found that the chi-square value of 2.020 had a probability between 0.10 and 0.20. This falls well inside the levels of 0.95 and 0.05. Therefore, a deviation from the ratio 2.0 to 1 as large as the one Mendel obtained gives no reason to reject the hypothesis that 2 to 1 is the true ratio. If Mendel had only known this, he never would have bothered to repeat the experiment. However, since he did repeat it, let's see how much he improved his situation.

The repeat experiment data included 65 plants with both green and yellow pods and 35 plants with only green pods. Since the total is still 100 plants, the expected values are still the same (Table E.3). The chi-square value, 0.1301, corresponds to a probability between 0.70 and 0.80. Here also there is no reason to reject the 2:1 hypothesis. It is true that the deviation is less here than before, but the essential judgment does not change.

Now let us go back to the argument that Mendel might have "cooked" his data to support an idea that he had had before he performed his experiments. What were the chi-square values for his monohybrid experiments? Do they support his conclusion that the ratio of dominating to recessive characters was indeed 3:1? The chi-square values for Mendel's seven monohybrid experiments are indicated in table E.4. The values of these chi-squares are all well below the 0.05 level of significance. Had Mendel known about and used the chi-square method, he would have concluded

Table E.3. Chi-square Test of Repeat of Experiment 5 of the Second Generation of Hybrids

Observed	Expected	$(O - E)$	$(O - E)^2$	$(O - E)^2/E$
65	66.7	−1.7	2.89	0.0433
35	33.7	+1.7	2.89	0.0868
				0.1301 = chi-square

Table E.4. Summary of Chi-squares for the Seven Monohybrid Experiments

Expected	Chi-square	df	Probability
1	0.262	1	$0.50 < p < 0.70$
2	0.015	1	$0.90 < p < 0.95$
3	0.390	1	$0.50 < p < 0.70$
4	0.063	1	$0.80 < p < 0.90$
5	0.449	1	$0.50 < p < 0.70$
6	0.349	1	$0.50 < p < 0.70$
7	0.606	1	$0.30 < p < 0.50$

that there was no reason to reject his hypothesis that among the offspring of his hybrids there were, on the average, 3 dominating to 1 recessive.

The same chi-square test has also been used to find out whether an experimenter has biased his or her data to meet a particular hypothesis. However, the appropriateness of its use for that purpose has been questioned. To understand how such a test can be used, suppose that the observed data in an experiment were identical to the expected data (the chi-square value would be zero). Obviously, you would be astonished and would likely be suspicious of how the experiment was run. Suppose that the observed data in another experiment were extremely close to the expected ones; for example, let us go back to Mendel's experiment 2. The chi-square of that experiment was found to be 0.015, a very low value indeed. Suppose that the value was even lower, say 0.00015. What are the chances that a deviation so small could occur just by chance? According to the table of chi-square, less than one percent; that is, a difference that small would occur less than once out of 100 times that the experiment was performed. Suppose that the same experiment was carried out a second time, again with a very low chi-square value. The probability that both experiments would have such a low value would be $0.01 \times 0.01 = 0.0001$.

Using the chi-square test in this fashion, Fisher came to the conclusion that Mendel's results were too good to be true. He assumed that the objective of Mendel's pea experiments was to verify a theory Mendel already had in mind. However, the authors of this interpretive text (Monaghan and Corcos 1985, 307) have demonstrated that the earliest time Mendel could have formulated a hypothesis to explain his data (his breeding ratios) in the course of his experiments was at the end of the second year of experiments 1 and 2, when he was dealing with seed traits. Had he known what to expect as the result of his first two pea-seed experiments, one might expect that the chi-square values of subsequent experiments dealing with mature plant traits ought to exhibit a trend toward smaller values. However, one cannot see any evidence of the expected trend (table E.4).

In a sophisticated paper, Franz Weiling (1986, 281) demonstrated that the conclusions drawn by Fisher were wrong because the assumptions on which his use of the chi-square test were based had not been met in Mendel's experiments. It looks as though Mendel has been vindicated, and yet in many biology textbooks the idea persists that Mendel might have been less than honest in reporting his data.

Glossary

Albumen	A nutritive substance surrounding a developing embryo. Mendel used this word interchangeably with endosperm. However, botanically there is no such thing as albumen or endosperm in the pea seed. What he was referring to was the cotyledons.
Amnion	A thin, tough membranous sac that contains a watery fluid in which the embryo of a mammal, bird, or reptile is suspended.
Annual	A plant in which the entire life cycle is completed in a single growing season.
Anther	The pollen-bearing portion of the stamen.
Anthocyanin	Any of a class of water-soluble pigments that impart to flowers and other plant parts any of the colors ranging from blue to most shades of red.
Asexual reproduction	Reproduction that does not involve the fusion of sex cells.
Bacteriophage	A submicroscopic, usually viral, organism that destroys bacteria.
Backcross	To mate a first-generation hybrid with one of its parents.
Breeding structure	Represents the combination of characters present in a plant as determined by observing the characters present in its progeny.
Calyx	Collective term for the sepals of a flower.
Carpel	A floral organ bearing and enclosing ovules; a pistil may be composed of one or more carpels.
Catechetics	Methods of teaching the basic religious doctrines of a church.

Character	A particular form of a trait; in Mendel's work, round or angular seed shape, yellow or green cotyledon color, terminal or axial flowers, tall or dwarf stem length, and so on.
Chromosome	Structural bodies, frequently rodlike, in the cell nucleus. They were shown in the twentieth century to be the sites of hereditary determiners, the genes.
Combination series	A sequence of all possible combinations of a group of objects taken two at a time, or three at a time, without regard of their order. For example, let us take the Ace and King of Spades from each of two decks of cards, giving us four cards. Taking two cards at a time, what possible pairs can we draw?—Ace-Ace, King-King, Ace-King, King-Ace. However, if we ask how many combinations of these cards are possible, the answer is only three, because two of the pairings, Ace-King and King-Ace, are the same.
Complete flower	One that has the usual flower parts (sepals, petals, stamens, and pistil).
Corolla	The petals of a flower taken collectively.
Cotyledons	Leaves of an embryo; also called seed leaves.
Cross-pollination	The transfer of pollen from the anther of one plant to the stigma of another.
Cyme	A usually broad, often flat, determinate type of inflorescence, i.e., with its central or terminate flowers blooming earliest.
Dihybrid	A cross (mating) between two individuals differing in two sets of contrasting characters.
Dioecious	Having male and female reproductive organs on separate plants.
DNA	Deoxyribose nucleic acid, the molecule found in chromosomes, shown to be one of the chemicals of heredity.
Dominating	A trait that passes into hybrid association unchanged is called dominating, and a trait that becomes latent into hybrid association is called recessive.
Embryo	The rudimentary plant developed from a zygote within an ovule.

Empirical	Relying upon or derived from observation or experiment.
Endosperm	A nutritive tissue surrounding the embryo.
Epigenesis	The theory that the individual is developed by structural elaboration of the unstructured egg rather than by a simple enlarging of a preformed entity.
Fertilization	The fusion of two sex cell nuclei.
Flavone	The parent substance of a number of important yellow pigments.
Gametes	Sex cells.
Gene	A functional hereditary unit, first imagined then found to be located on chromosomes.
Genetics	The modern study of heredity and variation based on the concept of genes.
Genotype	The genetic composition of an organism, determined by the assemblage of genes it possesses.
Grafting	A process by which a live part of a small branch of one variety of plant (the scion) is placed into a split stem or root of the same or a similar kind of plant (the stock). For the graft to be succesful, the growing layer of the scion must be placed in contact with the growing layer of the stock.
Germinal cells	Mendel's term for egg cells.
Heredity	The transmission of characteristics from parent to progeny.
Hybridization	Crossing of two plants that differ in one or more characteristic.
Hybrid vigor	The increased vitality of progeny, which frequently results when two different plants are crossed.
Hypocotyl	That part of the embryo between the radicle and the point of attachment of the cotyledons.
Hypothesis	A conjecture that accounts for a set of facts and that is subject to verification and, if verified, can be used as a basis for further investigation.
Imperfect flower	Flower lacking either stamen or pistil.
Incomplete flower	One that lacks one or more of the four kinds of flower parts.
Inflorescence	A cluster of flowers.
Interfertility	The ability of two different plants to produce offspring.

Keel	The two anterior petals of the pea flower
Law	A generalization or statement of a process that is not known to vary under a given set of conditions.
Mendelian genetics	This refers to four basic principles of genetics, which have been erroneously attributed to Mendel. These four principles are 1. *The principle of unit characters*. The inherited characteristics of an organism are controlled by factors, or genes, and these genes occur in pairs. 2. *The principle of dominance*. One gene in a pair may mask, hide, or inhibit the expression of the other gene. 3. *The principle of segregation*. The members of pairs of genes separate prior to gamete formation, and only one gene of each pair is present in a gamete. 4. *The principle of independent assortment*. The members of different pairs of genes are distributed to the gametes independently of one another.
Meiosis	Two special cell divisions occurring once in the life cycle of every sexual organism, plant or animal, halving the chromosome number.
Mitosis	Nuclear division in which the resultant nuclei in the daughter cells have the same number and kind of chromosomes as the original nucleus.
Monoecious	Having male and female sex structures borne on the same plant.
Monohybrid	A plant produced by crossing two varieties of a plant which differ in only one characteristic.
Monohybrid cross	A cross involving two plants that differ in one pair of contrasting characters.
Ovary	A structure located at the base of the pistil of a flower and containing the ovules of the plant. When the ovules are fertilized they develop into seeds.
Ovule	Structure in flowering plants, located inside the ovary, that eventually becomes the seed.
Perfect flower	One having both stamens and pistils

Petals	The colored, and therefore most conspicuous, parts of a flower. Usually regarded by the average person as the flower.
Phanerogram	An older term applied to a seed plant or flowering plant.
Phenotype	The external appearance of an organism.
Pistil	One of the main parts of a flower, usually consisting of stigma, style, and ovary.
Plumule	Another name for epicotyl.
Pod	The fruit of a plant, such as the pea, that contains several seeds; a pod usually dries up and splits open, liberating the seeds.
Pollination	The transfer of pollen from a stamen to a stigma by natural or artificial means.
Polyhybrid	A cross between two plants that differ by many pairs of characteristics.
Pomological society	Society whose members are interested in fruit trees.
Pomology	The study of fruits and fruit growing.
Preformation	A biological theory that assumes that all parts of a future organism exist completely formed in the sperm or the egg and that the development is simply by increase in size.
Raceme	Flowers born on an elongated axis on pedicels more or less equal in length.
Radicle	The root of an embryonic plant
Recessive	A term used to indicate a plant characteristic of one parent that is not expressed or shown in a hybrid but which may reappear in the offspring of the hybrid.
Recessive character	That characteristic whose appearance is excluded by the presence of its contrasting character in a hybrid.
Reciprocal crosses	Crosses in which the roles of the pollen and seed plants are reversed.
Reversion	Return to a former ancestral condition (also called, loosely, atavism).
Scion	*See* grafting
Seed coat	The outer layer of the seed, which develops from the integuments of the ovule.
Seed traits	Traits visible in the seed of the fertilized plant.

Selfing	The transfer by a plant breeder of pollen from an anther to the stigma on the same plant.
Self-pollination	The transfer of pollen from an anther to a stigma on the same plant.
Self-fertilization	Fertilization of a plant by its own pollen.
Sepal	Outermost flower structures, which usually enclose the other flower parts in the bud.
Shoot	A collective term for the stem and its leaves of a young plant.
Species	The smallest unit of taxonomic classification; a kind of plant or animal. The members of a species share similar structural characteristics, and all the members of the species have the potential to interbreed and to produce fertile offspring.
Stamen	The pollen-producing structure of the flower, consisting of an anther and a filament.
Standard	In a pea blossom, the standard is the petal that stands up vertically.
Stigma	The receptive part of a pistil, which receives the pollen and on which the pollen germinates.
Stock	*See* grafting
Style	Slender column of tissue that arises from the top of the ovary and through which the pollen tube grows.
Tarsi	The hind legs of an insect. In Mendel's paper, the hind legs of the *Bruchus pisi*.
Testcross	A cross made to determine the breeding structure of a plant. The plant whose breeding structure is unknown is crossed with a plant whose breeding structure is known to be recessive.
Trait	A general, distinguishing feature of an organism, in Mendel's work, seed shape, cotyledon color, seed coat color, position of flowers, stem length, and so on.
Trihybrid	A plant produced by crossing two varieties of a plant which differ in three characteristics.
True-breeding	A term used in plant breeding to indicate that the offspring have the same characteristics as their parents from generation to generation.
Trunk	The main stem of a tree.
Variety	A category within a species, a group of members distinguishable from the other groups within the

	species by the possession of a distinctive character or characters.
Vitalism	A theory of life that holds that life processes and phenomena are the results of forces that exist in addition to those forces that are physical and chemical in nature. The opposing view is *mechanism*.
Viticulture	The science of growing grapes.
Wings	The two lateral petals of a pea flower.
Zygote	A fertilized egg, the first cell of an organism.

Bibliography

Bateson, B. 1928. *William Bateson: F.R.S. Naturalist.* London: Cambridge University Press.

Berrie, A. 1977. *An Introduction to the Botany of the Major Crop Plants.* London: Heyden.

Briggs, F. N. and P. F. Knowles. 1967. *Introduction to Plant Breeding.* Reinhold Books in Agricultural Science. New York: Reinhold.

Bronowski, J. 1973. *The Ascent of Man.* Boston: Little, Brown.

Corcos, A., and F. Monaghan. 1986. "More about Mendel's Experiments: Where Is the Bias?" *Journal of Heredity* 76:384.

Hayes, H. K., and R. J. Garber. 1921. *Breeding Crop Plants.* New York: McGraw-Hill.

Heimans, J. 1970. "A Recently Discovered Note on Hybridization in Mendel's Handwriting." *Folia Mendeliana* 5:13–24.

Iltis, A. 1954. "Gregor Mendel's Autobiography." *Journal of Heredity* 45:231–234.

Iltis, H. [1932] 1966. *Life of Mendel.* Translated by Eden and Cedar Paul. New York: W. W. Norton. Reprint, with 10 illustrations and 12 plates. New York: Hafner.

Keller-Fox, E. 1983. *A Feeling for the Organism. The Life and Work of Barbara McClintock.* New York: W. H. Freeman.

Matalova, A. 1984. "Response to Mendel's Death in 1884." *Folia Mendeliana* 19:212–221.

Matalova, A., and A. Kabelka. 1982. "The Beehouse of Gregor Mendel." *Folia Mendeliana* 17:207–212.

Messenger. March 1938–November 1941. "Gregor Mendel: Abbot and Discoverer of the Laws of Heredity." Biography serialized in the journal *The Messenger,* published by the Diocese of Covington, Kentucky.

Monaghan F., and A. Corcos. 1985. "Chi-square and Mendel's Experiments: Where Is the Bias?" *Journal of Heredity* 76:307–309.

Olby, R. 1985. *The Origins of Mendelism.* 2d ed. Chicago: University of Chicago Press.

Orel, V. 1970. "Mendel and the Central Board of the Agricultural Society." *Folia Mendeliana* 5:39–54.

Orel, V. 1984a. *Mendel.* New York: Oxford University Press.

Orel, V. 1984b. "Mendel's Involvement in the Plea for Freedom of Teaching in the Revolutionary War of 1848." *Folia Mendeliana* 19:223–233.

Romanis, A. G. de 1929. "A Great Innovator in the Study of Natural Sciences: F. John Gregory Mendel, Augustinian." *Augustinian Historical Bulletin,* November 13, 1929. Florence, Italy.

Stern, C., and E. Sherwood, eds. 1966. *The Origin of Genetics: A Mendel Source Book.* San Francisco: W. H. Freeman.

Weiling, F. 1986. "What about R. A. Fisher's Statement of the 'Too Good' Data of J. G. Mendel's Paper?" *Journal of Heredity* 77:281–283.

Zumkeller, Adolar. 1971. "Recently Discovered Sermon Sketches of Gregor Mendel." *Folia Mendeliana* 6:247–252.

Index